Field Adventures

in Paleontology

Lynne M. Clos

Fossil News
Boulder, Colorado

This book was produced entirely on Macintosh computers—an iBook, a G4 Cube, and a G3 Desktop. The text was set in 12 pt. Hoefler Text and the headings are in 30 pt. Carlton LET. Adobe Pagemaker software was used for the layout.

Book and cover design by Lynne M. Clos

Printed in U.S.A.

ISBN 0-9724416-3-8

Library of Congress Control Number: 2003094124

To Norm—without those long lunchtime
talks and the inspiration of your example,
I might never have had the wisdom
to follow my heart.

Contents

Foreword

Hey! Who's that cool-looking blonde behind those Foster Grants?—I can still hear myself thinking about six years ago after opening an order of *Fossil News* magazine back issues. I was looking at page 7 in the August 1995 issue of (what was then) Joe Small's *Fossil News* magazine. The name 'Clos' kept popping up in issue after issue, article after article—including a great feature documenting her trip to Alaska's North Slope in search of dinosaur bones. "Very cool!" I thought, and that wasn't just a reference to the climate there. "This paleontology lover really writes her *** off," I mused! With retro-apologies for my chauvinistic tone, this was my introduction to a remarkable person I would become well acquainted with through email correspondence in time. (Later I discovered that at roughly the same 'era,' we honed our early interests in prehistory on the same children's book, Darlene Geis' *Dinosaurs and Other Prehistoric Animals*, 1959.)

Ever since, I've admired Lynne for accomplishing things in her life with utmost dedication that I considered doing, but never had the temerity to try. Like going back to school, as she did in her mid-30s for a coveted graduate degree in paleontology/museum studies, and writing a professional scientific paper on an extinct vertebrate, like she did for her thesis on a Lower Miocene varanid reptile *(Varanus rusingensis)*. Or for throwing caution to the wind in pursuit of fossils found in remote and exotic places. And, ever since the fall of 1998, how does she somehow find the 'extra' time to publish the world's foremost popular monthly magazine exploring the lure of fossil collecting, the lore of prehistory and the love of paleontology? Given those credentials, naturally I was elated when she took me aboard as one of *Fossil News'* regular contributing editors.

Scores of books have been published on dinosaurs, and others offering tips on how to collect fossils (usually invertebrates). But few have a distinctive autobiographical 'flavor' while emphasizing the 'romancing' of old stones and bones, perhaps the inspiration behind Gideon Mantell's popular book *Wonders of Geology,* written over 160 years ago. Two more recent examples which come to mind are Edwin H. Colbert's *A Fossil Hunter's Notebook* (1980), and Michael Novacek's *Time Traveler* (2002), both of which provided many hours of enjoyable armchair experiences to reflect upon, vicariously!   This is the tradition into which Lynne's new entry, *Field Adventures in Paleontology,* falls.

Okay, get it straight. Lynne Clos is no Roy Chapman Andrews 'swashbuckler' type, yet she's dug fossils in places that even Andrews would have heartily agreed qualify as 'the ends of the earth.'   Another refreshing aspect to Lynne's work is that, historically, few females do this particular kind of natural history writing. One such example is Zofia Kielan-Jaworowska's *Hunting for Dinosaurs* (MIT Press, 1969) concerning the successes of the Polish-Mongolian team, which later, in 1971, discovered the famous 'fighting dinosaurs' pair—*Velociraptor* and *Protoceratops.*   From what I've read in her own *Field Adventures,* had she been a bit older then and accessed the right international connections, Lynne would have made us proud representing the United States on that expedition. One wonders what "wonders" she would have unearthed there?

The beauty of it all, though, is that every 'bug hunter' has a story to tell about fossils—relics from the past which have left their impressions in the rocks, as well as indelibly in our hearts and minds—which is why I read cycad-loving Lynne's tales with enthusiasm. Besides, I know from reading her material that she's a wonderful teacher. Someday, before my pending day of fossilization, I hope to don a pair of Foster Grants and head off yonder with her in search of fossils big and small, recent and most ancient. But beware, Lynne—for like one of your acquaintances, I too have a "mag-gnat-ic" personality!

True—we'd learn lots collecting fossils with Lynne and it would be great, memorable fun. But for now we can enjoy her experiences vicariously. So all ye fossil collectors out there—settle back in your easy chairs. Lynne has some interesting things to relate about her past and deep time (not that she's that old, of course). Enjoy and get inspired, like I did. And don't be surprised if maybe some of you find yourselves registering for a paleo-course somewhere down the line too...

Allen A. Debus
Hanover Park, IL
May, 2003

Preface

I have always been interested in prehistoric life. My dad had a set of plastic dinosaurs on a shelf in his workshop way back when we lived in Baton Rouge, during my preschool years. Somewhere along in that time I acquired a small grey *Allosaurus* of my own, which I still have. Although I grew up in a scientific family, the daughter of a physics professor and an occupational therapist-turned-stay-at-home-mom, neither of my parents had any particular expertise in paleontology.

When I was four, my older sister was learning to read. She had what was then called a "learning disability," and reading was a real challenge for her. My parents would patiently, endlessly work with her night after night, while I sat up on the back of our couch, looking on and soaking it all in. I soon tired of Dick and Jane, and wanted some books of my own. One time, my mom took us to a bookstore. I was too short to see most of the books way up on the shelves, but they had a sale table that was just at my height. On it I spied a large book with a wonderful painting of an *Allosaurus* on the dust jacket. I refused to leave the store without that book. It was *Dinosaurs and Other Prehistoric Animals* by Darlene Geis, really written for children two or three times my age—but that didn't intimidate me, and with help from both the pronunciation guide and my parents, I really learned to read from that book. I still have it, although its age and the amount of use it got are quite apparent.

That book opened my eyes onto worlds very different from our own. Trilobites swam in tropical seas, Mary Anning found an ichthyosaur when she was still a girl, and *Triceratops* battled *Tyrannosaurus rex*. In addition to reading and rereading those tales, I copied the pictures with my crayons and assembled my own little "dinosaur book." I had to ask my mom how to spell "cute" when I wrote the caption for *Protoceratops*, and she questioned whether I would always feel that way towards what some would consider an overgrown, warty reptile; yes, mom, by this time you know I do…witness that ceramic sculpture of a Proto hatching out of an egg that sits on a shelf in my office! We moved to the northern suburbs of Detroit when I was about to enter the first grade, and unfortunately, there were no museums nearby at that time in which I could see actual dinosaur skeletons. The Cranbrook Institute of Science had a slab of trilobites and some dioramas, but for the most part, I viewed prehistoric worlds through my books.

In those days, I was the only kid on the school bus who knew what an *Archaeopteryx* was, much less could pronounce it. I didn't even know anyone else who would *want* to know! Fossil-hunting clubs were nowhere to be found, especially in a part of the country where everything is covered with glacial till. Once they thought they had found a mammoth skeleton north of town, only to discover that a passing circus had buried a dead elephant there on the sly.

Needless to say, my goal of becoming a paleontologist was not a popular one. When you grow up near Detroit, and are good at math, you are *supposed* to become an automotive engineer. It's what everybody does. Paleontology? You've got to be kidding! Grow up, get a real job. Not having the resources to go to the University of Michigan, the only place in the state with a paleo program and my first choice of colleges, I listened to my elders and earned a bachelor's degree in mechanical engineering, and went to work for General Motors. After all, I'd never been on a dig in my life, so what did I know about what real paleontology was like?

My job at the GM Proving Grounds wasn't a bad one, but neither did it inspire me. Nobody could believe I would give up a "good job" just to relocate out west, but I was tired of southern Michigan, where for six months out of the year the sky is grey, the trees are grey, and even the snow is grey. During my sophomore year of college, I had quit for a while and lived in New Mexico making turquoise jewelry for a living, and had fallen in love with the west. So I took another engineering job in Colorado, and the corporate politics there made the Proving Grounds look like a country club. After eight years in engineering, I knew I had made a mistake. What had happened to those dreams of digging for dinosaur bones? If I tried it, would I be as disillusioned as I was with a corporate job?

Well, there was only one way to find out. In 1986, I signed up with an Earthwatch expedition to dig for small dinosaurs on Colorado's Western Slope. I thought maybe I'd hate it, and then I'd get those nagging thoughts of wanting to be a paleontologist out of my mind.

Instead, I was in heaven. No reams of government paperwork to fill out, no arguing with people with a "that's not my job, man" attitude, no unrealistic production schedules...just fabulous scenery, intelligent and interesting companions, the stillness of the high desert, and the wonder of setting eyes on bone fragments (or skeletons!) that hadn't seen the light of day for over 140 million years. Of course, there wasn't much money in it, but what good is money if you hate going to work every morning and feel like a caged animal while you're there? By this time, I had put my husband through grad school and bought a house, so I figured it was my turn. I came home from that dig and told Chris I was going back to school to study paleontology.

Earning my Master's degree in museum studies, with a specialization in vertebrate paleontology, took a lot of dedication over the next five years. I have suffered from frequent migraines all my life, and the problem has only become worse as I've gotten older. That in itself made grad school difficult, but add to it a difficult pregnancy and a new baby, along with the demands of changing fields, and it was quite a challenge. Had my situation been different, I would have loved to continue through a Ph.D. But as it was, I decided a Master's was enough, and finished in 1991 (although I missed my graduation ceremony because of an opportunity to visit Japan).

Any of you who are mothers will understand that my two daughters come first in my life. Sometimes I could really use a wife! But a person has to live within her limitations, and between my family and my headaches, I could not see a tenure-track professorship or museum curatorship in my future. So when I went on digs I did so as a volunteer, and once my kids were in school, I took over as editor and publisher of the journal *Fossil News*. I've always liked to write (remember that little "dinosaur book"?), and running a home business gives me the opportunity to do paleontology while flexing around my kids' needs and my headaches. I love it, and it works for me.

Early in 1998 I decided I wanted to create a website. I was amazed to see that among all the websites dedicated to paleo topics, none dealt with what it is really like to do paleontology in the field. Well, by this time, I'd been on quite a few digs, so I figured—why not? I don't need management approval to do it myself! Over the years, I have received many emails from people who are delighted to read real accounts of fossil expeditions, including students seeking advice on real-world paleontology.

Hence this book. You can pick up any number of volumes by Ph.D.'s that include some material about field paleontology, but I don't know of another that covers the variety of fossils, techniques, time periods, and places that I have included here. If you are an armchair paleontology fan who wonders what fieldwork is like, a seasoned collector who's never participated in a professional dig, or a youngster considering a career in the field, I've written this book for you!

Acknowledgments

Many people were instrumental in making this book what it is. Although the writing, digital artwork, most of the photography, and the book design and layout are entirely my own, I wouldn't have had anything to write about were it not for the contributions of a great many friends, colleagues, and mentors.

First and foremost, I appreciate the support of my family. My mom, Teddie Mobley, has accompanied me on several of these digs, and took the pictures that I did not. My dad, Ralph, isn't really into hunting fossils, but has watched my kids so that my mom and I could dig together. My parents watched the kids when they were little so that I could go on some of the trips to faraway places. My daughters, Allison and Mattie, have been great little fossil-hunting companions. And finally, without my husband Chris' willingness to be the sole support of the family for many years, I would not have been able to escape a lucrative, boring job for a rewarding career that yields only a pittance. And he did let me drag him out to the trilobite beds in the Utah desert on our honeymoon.

I have benefited also from the guidance of many people within both the amateur and the professional paleontological communities. Some of these folks shared their knowledge of fossils or geology with me; others showed me where to collect or ran the digs I went on. Without them, I would not have had the opportunity to participate in these digs, let alone write a book about them. Whereas a complete list of all these people is impossible to compile, I particularly appreciate the contributions of George Callison, Kirk Johnson, Lou Taylor, Karen Houck, Roland Gangloff, Toni Superchi, Don Rasmussen, Mike Parrish, Vince Santucci, Herb Meyer, Tony Barnosky,

Malcolm Bedell, Jordan Sawdo, Wayne Itano, Chris Weege, Fred Olsen, and Doug Nelson. Allen Debus has written a terrific fore-word. Had I not had the opportunity to become a member of the Western Interior Paleontological Society, many of these digs would not have come about, and there are members of that organization too numerous to name who have been great digging companions over the years.

Finally, I wish to acknowledge the inspiration provided by my dear friend Norm James. He left a cushy job with GM at age 35 to go back to grad school and change careers from chemistry to exercise physiology, where his heart really lay. When I was contemplating my own career change, I often thought of Norm, and told myself, "If he could do it, so can I."

A Short Stop at a Kentucky Roadcut

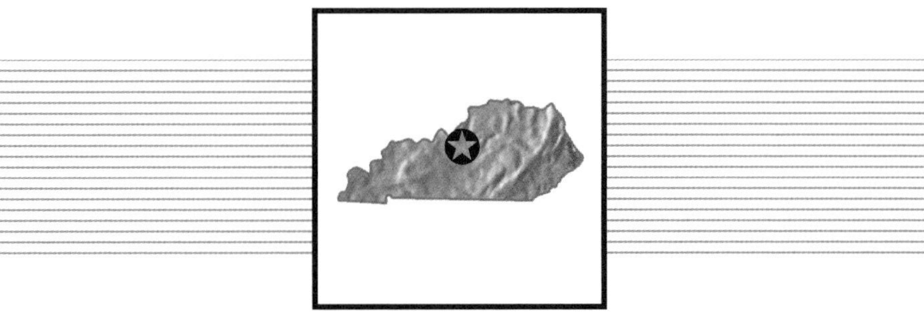

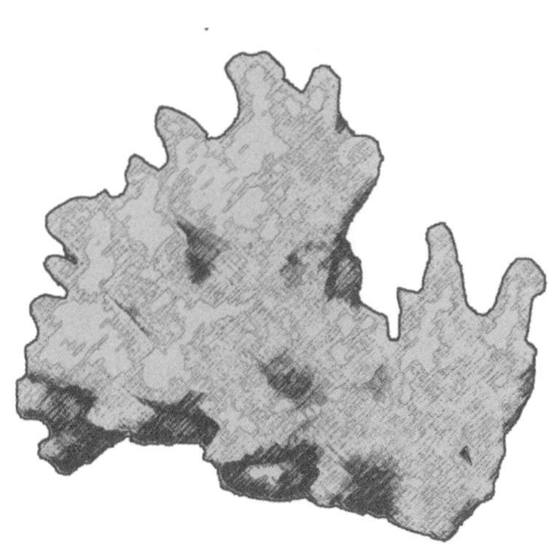

I have always been interested in fossils since I was a kid, but when I was little we lived on the Mississippi Delta in Louisiana, and once I finished kindergarten, we moved to southern Michigan, where everything is covered with glacial till—nary an outcrop, let alone a fossiliferous one, to be found anywhere. I did plenty of reading about prehistoric life while I was growing up, but by the time I was 22, I had yet to collect my first fossil. Growing up north of Detroit, everybody (and I mean *everybody*) who is a good student—especially in math—is encouraged, even leaned on, to pursue a career in the automotive industry, so of course my college studies were in engineering. There were no paleontology (or even geology) offerings at the university I attended, where I might have met up with like-minded folks. So my first "collecting trip" was pretty inauspicious, but it was memorable nonetheless.

Split rail fenceline in Kentucky.

In April of 1978, my Aunt Dotli (my mom's twin sister) was visiting from California, and we decided to drive south and visit Mammoth Cave in Kentucky, as we had always wanted to go there. I had seen a couple of smaller, decorated caves in Virginia a few years before, when my dad attended a conference in Washington, D.C. and my sister and I tagged along, but I had long been intrigued by the "big two"—Mammoth Cave and Carlsbad Caverns—and fell in love with Carlsbad when I was returning from a stint as a college dropout in New Mexico, so I couldn't wait to see Mammoth. I love travelling, too—as you'll discover as you read through this book—and am always up for something new. One of these years I'll get back to Carlsbad in June to see the bat flights at dusk, and now that I live in Colorado, the trek won't be as far.

Flowstone decorations in Mammoth Cave.

I've always been the one in our family who plans the itinerary. Even when I was around 10 years old, and we'd take one of our every-few-years trips to Wisconsin to see my gramma, I was the one who'd pull out the maps and travel guides and see what cool places we might stop along the way. (Once we even came across a small museum in the northern Lower Peninsula, Haltiner's Hall of Ancient Man, where we got to see some real shrunken heads from South America.) So, as usual, I plotted our route to Mammoth Cave, detouring past an in-

These small fossils were the first ones I collected myself.

teresting-sounding state park called Big Bone Lick, which turned out to have been a Pleistocene sinkhole where mammoths and other fauna became trapped, with a small museum displaying the fossils.

The cave was fabulous—the tour took all day, and I couldn't believe all the underground rivers and lakes, something you don't see at Carlsbad. Stalactites, stalagmites, and other cave decorations adorned every square inch of the walls in most places. Of course, it is not only illegal but unethical to break off cave formations there, but we later stopped at a roadside stand where my sister bought a small stalactite someone had collected somewhere else. I suppose in retrospect that by buying it we were perpetuating the destruction of priceless geological treasures, but having no clue where it came from— it might have been private property—we didn't worry about it too much at the time.

Kentucky was beyond the reach of Ice Age glaciers, so we passed plenty of outcrops as we drove along the roads there. I remember a lot of red clay soil that stood in stark contrast to the thick, verdant forest. I'm not sure why we decided to stop at one spot and look for fossils; maybe the same guy who sold my sister the stalactite had had other geological artifacts for sale—I really can't remember. But when we did stop, all we had to do was climb the embankment a little and the ground was littered with fossil corals. I haven't a clue what formation or age we were in, but judging from the specimens we found, it was likely pre-Tertiary. I found a nice chunk of a reddish-hued tabulate coral, some smaller fossils that look more like Coolie hats than anything else, and a few fragments of branching corals. At the time, I was so excited to be collecting my own fossils that I really didn't worry about the fact that I knew next to nothing about them. To this day, I keep them in a cigar box and show them, along with various other fossils, to the kids when I give a fossil talk at a school.

So that was the inauspicious beginning of my field career. Later that summer, my parents, sister, and I would make our first trip to Utah, where we came to collect some fabulous trilobite beds that put the Kentucky corals to shame...but I'll save that for a later chapter.

2

Fruita Paleontological Area

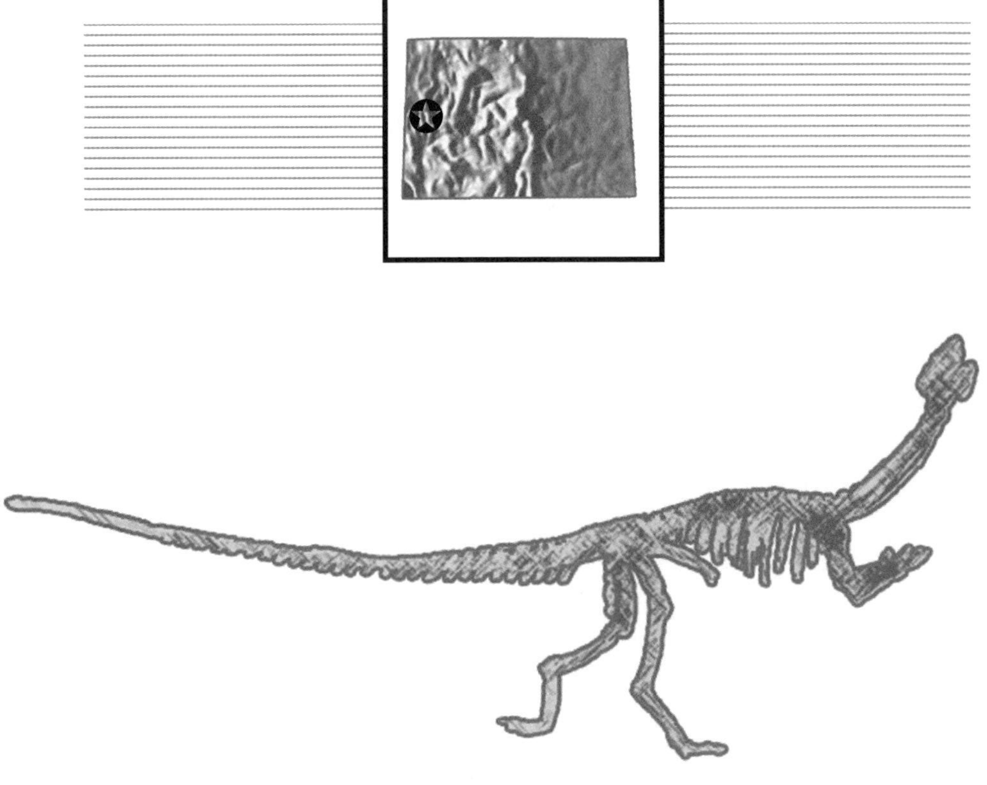

The first serious scientific dig I participated in was an Earthwatch expedition aimed at finding tiny dinosaurs and other small fossils in the badlands of western Colorado. At the time, I was working as a mechanical engineer for Rockwell International, and hated my job. I couldn't wait for the opportunity to pursue the interest in paleontology I'd had since I read my first dinosaur book at age 4. Oh, I'd done a bit of personal collecting of invertebrates over the years, but I'd never done any real science.

Beth and Bob digging at one of several quarries at Fruita.

The dig was led by George Callison of California State University at Long Beach. George is a great guy and a lot of fun to dig with. The dig site is a specially protected area on Bureau of Land Management (BLM) land just outside Fruita, a small town west of Grand Junction. In the years since my dig with George, Fruita has been put on the map by the establishment of the Museum of Western Colorado's Dinosaur Journey, which is a small but fascinating museum with fossils and large robotic dinosaurs. However, in July of 1986, Fruita was just one of many sleepy little towns on Colorado's Western Slope.

The Coke Ovens at Colorado National Monument.

The Fruita Paleontological Area (FPA) is situated in the Grand Valley, an easy walk from the southern bank of the Colorado River. The Grand Valley is ringed by the Book Cliffs to the north, the Uncompahgre Plateau to the south, and the Grand Mesa to the east. The sediments of the valley floor in the area of the FPA are soft grey shales of the Brushy Basin Member of the Jurassic Morrison Formation, about 144 million years old. The sediments dip to the north, so that as you drive south and up onto the rimrock where Colorado National Monument is located (our base camp), you go through older rocks. The Book Cliffs on the north consist of exposures of Tertiary sediments. The Grand Mesa is a high plateau capped by volcanic rocks.

The Morrison Formation is famous for its dinosaurs, but most of them are the familiar giants like *Apatosaurus* ("Brontosaurus") and *Stegosaurus*, the Colorado state fossil. The formation was named for the tiny town of Morrison, outside Denver, more than a century ago, when large dinosaur bones were discovered there on what is now known as Dinosaur Ridge. The Morrison is widely exposed in the western states and consists of a series of sandstones, shales, and lacustrine (freshwater lake) limestones deposited on a low coastal plain during Late Jurassic time. Once you've seen the Morrison, you can recognize it practically anywhere, by the distinctive juxtaposition of beds that are purplish-red, grey, and grey-green in color.

Rich and George at Tom's Place. The hill in the foreground shows the "popcorn" texture of weathered bentonite.

The sediments of the Morrison exposed at the FPA are grey bentonitic shales formed from the compaction and weathering of volcanic ash. It is thought that the volcanic ash was redistributed by streams as fluvial and overbank deposits. The rock is extremely fine-grained, and is one of the few places in the Morrison (or anywhere) to preserve the delicate bones of tiny dinosaurs, lizards, and mammals. The shale weathers on the surface to the "popcorn" typical of bentonite, destroying any delicate fossils in the process; so the overburden needs to be removed first, and the virgin, unweathered rock exposed, before prospecting can begin.

But before we got down to any serious digging, George took us on a hike to acquaint us with the area. Up hillocks, down arroyos, and past crunchy sagebrush we trod, and every so often George would stop, pick something up, and say, "This is a piece of dinosaur bone...oh, and here's another. But these are from big dinosaurs. We're not interested in those. We want the little guys."

How could he be so nonchalant about dinosaur bones? This was the most exciting thing I'd ever done! And how could he tell that

Would you know these were fragments of dinosaur bone at first glance?

these little grey, purplish, and red-dish rocks were pieces of dinosaur bone? They just looked like rocks to me!

Never fear, George assured us. Look at the texture. It doesn't look much like these other rocks, does it? It looks more like the spongy texture of this broken, weathered cow vertebra here...see? Once you develop an "eye" for it, it will just jump right out at you. He was right, and by the end of the hike I was spotting my own dinosaur bone fragments, obvious (now) against the light grey clay.

George and his crew had been working the area in previous years, and in the main quarry area, the overburden had been removed by bulldozer and a grid staked out. When my team was assembled and ready to begin, visqueen laid at the end of the previous field season to protect the pit over the winter had to be removed first. In addition to the main pit, past surface surveys had indicated several other promising sites within walking distance, and George assigned some of us to dig at those; I was sent to the top of a small ridge closer to the river, to a locality designated "Tom's place." I spent the entire two weeks there, digging slowly, inspecting each piece of rock for bone, and throwing the discards down the hill. A fortnight of excavation by three of us created a ledge about six feet wide, a dozen feet long, and waist-deep.

George showed us all on the first day what the bones he was interested in looked like. They resembled pencil leads more than anything else. The rock was light grey, and the bone was black; these bones are so delicate and small that they are not removed from matrix in the field. When any of us found a piece of rock with something black in it, we called one of George's field assistants over and he or she inspected it with a hand lens to determine if the discoloration was bone, a manganese stain, a root cast, or something else of little interest. If it was bone, a preservative was applied and allowed to dry, and then the block was carefully wrapped in toilet paper, sealed with masking tape, and labeled. Preparation of the tiny bones had to be done later in the lab, using fine tools and microscopes.

We found numerous "keepers" every day, but most of them didn't look like much in field condition: longbones in cross-section, things like that. (Longbone is a loose term covering any of the long, thin bones of the arm or leg: humerus, tibia, femur, etc.) They couldn't be identified as to element let alone animal until after lab preparation was completed. Nevertheless, each find was exciting, because we could imagine that a splendid fossil might eventually emerge. Digging was tedious, but fun, despite the hot (90°+) sun and relentless aggravation of biting gnats. At least I did not have as much trouble with gnat bites as some. For one thing, the gnats took cover during the hottest part of the day, leaving us only the heat to contend with. For another, they seemed to prefer some people over others. Gordon, in particular, had a "mag-gnat-ic" personality, and the citronella oil he applied in attempted retaliation seemed to do little other than scent our quarry ledge. And some people had an allergy to the bites; my

Gordon spent the entire two weeks excavating his little "cave" and perfuming our quarry ledge with citronella oil.

tent-mate, Winona, got welts the size of fifty-cent pieces wherever she had been bitten. I was annoyed to have little chunks of my flesh removed, and to have the beasties dive-bomb my eyes the way gnats always do, but other than that, I didn't have a real problem.

George didn't work us dawn to dusk seven days a week. We did start quite early to avoid the heat, but knocked off by mid-afternoon. Some people took a dip in the Colorado River at lunchtime to cool off, although you had to be careful to stay close to shore where the current wasn't too strong. I only cooled my feet in the river, but one day when it was really hot and both Winona and I were feeling rather ill from the heat, we drove into town and swam in the public pool (it was convenient that I had driven my own car to the dig). By late afternoon, we were back up at the Monument, which was a couple thousand feet higher in elevation and shaded by piñon and juniper trees, so it was much more comfortable. It cooled off so much at night that we even had campfires. Every few days, we would go on a "field trip" to some interesting location in the area (and there are plenty of them, geologically and paleontologically speaking). One day we visited Rabbit Valley in the morning—a dinosaur bone interpretive trail a short hop down the freeway—and then continued on up to Douglas Pass, where we looked for fossil plants and insects in the exposures of Eocene Green River Shale right by the roadside. Another day we took a ride to the Dry Mesa quarry west of Delta, where fluvial (river-deposited) Morrison sandstones have yielded the big dinosaur bones, including

Rich taking a break just before he made the "big find."

the remains of "Ultrasaurus." This was really fascinating, although the ride in to the quarry was bumpy and long. I had never seen whole, big dinosaur bones being dug out of the ground—just pictures in books and skeletons in museums. It was so different from the quarrying for small fry that we were doing! On Sunday we skipped paleontology completely and went swimming at "The Potholes" on the Little Dolores River. These natural potholes form at the foot of a waterfall as the pounding waters hit loose rocks and knock them around like grinding stones. They make great swimming holes in a river that otherwise would be too shallow. The next day it was back to work, refreshed, at the Fruita quarry.

Aren't the best things always saved for last? On our final day of digging at Tom's place, while we were taking our lunch break, Rich was halfheartedly knocking away at some unpromising-looking blocks. All of a sudden, one split open to reveal the complete, articulated skeleton of a small lizard! We couldn't believe our eyes! I rushed to photograph it before it was preserved and wrapped, something that

Part and counterpart of the lizard skeleton.

had to be done quickly since the virgin rock is slightly damp and will crack as it dries if the preservative isn't applied to bind it together. We ran to tell George, who immediately came to have a look, and declared that this was by far the best-preserved small lizard from the Late Jurassic yet to be found in North America. It made two weeks of finding nothing but pencil leads worthwhile.

The remainder of the afternoon, of course, was anticlimactic. Now that we were sure there really were fossils in those rocks (and didn't have to take the professionals' word for it that black specks were bone), we kept thinking that where there was one articulated skeleton, there should be more...but they remained hidden, waiting for the next crew to find.

On the drive back to Boulder, I formulated just how I would tell my husband that I was going to quit my job and go back to college to study paleontology. When I had signed up for the dig, I thought maybe I would hate it, and then I would get these nagging thoughts about wanting to do paleontology instead of engineering out of my mind. Instead, I had a blast, and I knew I couldn't spend the next 30 years of my life chained to a corporate job. To my amaze-ment, Chris didn't give me a hard time about my announcement; he said he would be glad not to have to hear me complain about work any more. So, at age 31, that's what I did. I already lived in Boulder, so the University of Colorado couldn't have been more convenient. It took me five years to finish a Master's degree in museum studies with an emphasis on vertebrate paleontology, because I had to start out making up a lot of undergraduate classes in biology and geology, and I also had a baby along the way (a delightful little girl). But I truly enjoyed school for the first time in my life, and it was one of the best decisions I've ever made.

One of the locals, an inquisitive collared lizard.

3

Pawnee Buttes Anthills

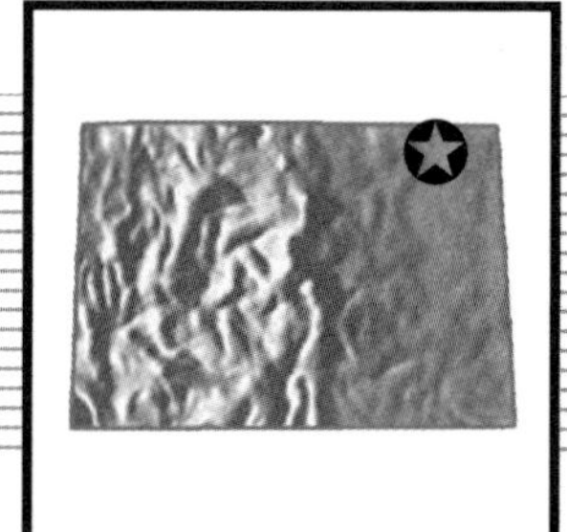

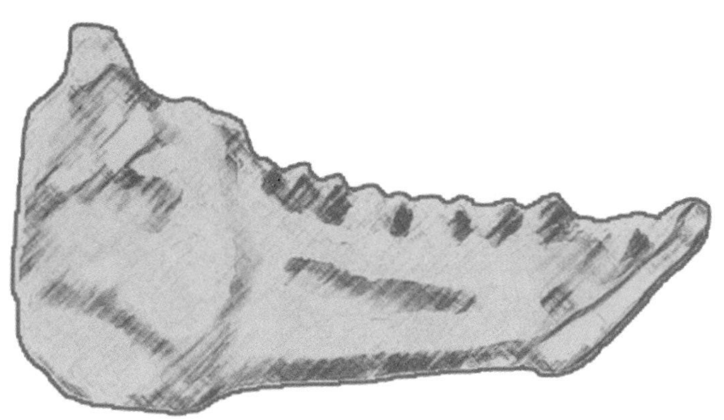

Once I had decided to get out of engineering and go back to school to study paleontology, I couldn't wait to get started. I made the decision in July, which didn't give me enough time to get all the paperwork together and apply as a full-time student for the fall semester; so I decided to continue working until January, and take one introductory geology course at night, and then

Prairie landscape with Pawnee Buttes in the distance.

start full-time in school for the winter semester. My lab instructor, Karen Houck, recommended that we go to the Denver Gem & Mineral Show in September and look around; so I did, and it was there that I stumbled across an unattended table with information on it announcing the recent formation of the Western Interior Paleontological Society (WIPS), an amateur group in the area. I took a membership application home with me, and attended my first meeting of the fledgling group in November of 1986 at the Denver Museum of Natural History. I've been a member ever since.

I had never had the opportunity to belong to a group of other fossil enthusiasts. In fact, I'd never met anyone else on the planet who thought fossils were totally cool, besides my mom and the other participants in the Fruita dig. Every month there was a lecture on some facet of paleontology, and when spring rolled around, there were field trips a person could sign up for. Now maybe I'd find out where the good outcrops were around here!

My first field trip, in March of 1987, was to the Pawnee National Grassland in northeastern Colorado, to collect anthills in the Oligocene White River Formation. Anthills, you say? Well, apparently some kinds of ants like to build their hills out of interesting small particles they find, and they frequently include microfossils among their building materials. The trip leader, Lou Taylor, explained to us that all we had to do was find an anthill, scoop some material into a ziploc without including too many irate ants, and then take it home, screen out the fine dust, and sort through what was left using a hand lens or microscope, and we were likely to find rodent teeth, small bones, ostracods, conodonts, eggshell fragments, and all sorts of wondrous little treasures.

I'd read lots of books about prehistoric life over the years, but most of them contained precious little about fieldwork, and I had certainly never heard of letting ants do the work for you! Still, I thirsted for every new opportunity I could get my hands on. So I loaded up some gear, water, and my faithful little black mutt Rooter into my 74 Gremlin, and headed out I-76 to the rendezvous point. From there we would caravan in to the area Lou had in mind.

The Pawnee Buttes are long sandstone bluffs that stand out for miles against the prairie landscape. The White River Group is a terrestrial unit that represents river and floodplain deposits formed during a time when grasslands were just starting to appear and spread across the North American continent. The Cretaceous epicontinental seas had dried up, and the eastern and western halves of North America were united to form pretty much the continent as we now know it. The dinosaurs were gone, the Rockies had risen, and the vast interior plains were populated by a fantastic menagerie of brontotheres, running rhinos, small camels, antelope-like creatures, and huge herds of oreodonts, little artiodactyls that resembled small, slender sheep—and of course, a bunch of tiny critters that lived beneath their feet. The area we were going to was in the Horsetail Creek Member of the White River, which correlates to the Chadronian Land Mammal Age, about 30-34 million years ago. The Horsetail Creek lies unconformably atop Cretaceous sandstones—meaning that there is a gap in the rock record corresponding to the timespan between the Cretaceous and the Oligocene, representing a time either when no sediments were deposited, or they were worn away before the White River was laid down. The White River is a very famous unit which has produced scads of

Typical fossil-bearing anthill.

fossils from the Colorado-Nebraska-South Dakota area, and is exposed in Badlands National Park.

As Lou explained to us before setting us loose, there is a particular method to collecting anthills. You don't just dig up the whole hill—it isn't necessary, and it isn't very nice to the ants. An anthill left standing may produce a new crop of microfossils year after year. All you need to do is use a trowel to scrape off the large-particled layer that lies atop the fine dust that makes up most of the hill. That way the ants' home isn't disturbed too much, and that's where all the fossils are anyway.

Not every anthill will produce fossils—it depends on whether any loose microfossils have weathered out within the ants' home range—so it's worth looking closely at the outer layer of the hill first, to see if you can spot any odd particles that might be fossils. White River fossils tend to be whitish in color, whereas most of the outer layer of the anthill is made up of small gravel and dust clumps that are more light brownish-grey in color. (It looks a lot like kitty litter.) So an anthill that contains obvious white particles mixed in with the brown is more promising than one that doesn't. Good eyesight (which I had back then!) helps, though, because many of the white particles may merely be grains of quartz.

So while I tromped around the prairie, avoiding prickly pears and yuccas and looking for anthills, Rooter roamed the open spaces too, finding all sorts of exciting things to sniff. I swear that dog could smell in color. He wasn't very good at finding fossils, but we both had a great time anyway. It was amazing how many anthills there were out in those grasslands—enough for all dozen or so of us to find several of our own. Big anthills, too: sometimes half a foot high and a foot or more in diameter. I didn't collect every one I came across, because I knew I'd never find the time to sort through all that material. So by midafternoon I had only bagged up material from the most promising-looking two.

Back home, the material needed to be screened and washed. I have a couple of large, round screens (a little bigger than cake pans) of different mesh sizes that my mom gave to me when she gave me her kiln and all her ceramics supplies. I'm not sure if they were geological paraphernalia that she co-opted for screening glaze chemicals or the other way around, but they work great. I just dumped a baggie of anthill material onto a fine-mesh screen, put it in the kitchen sink, and sprayed it all over with the sprayer hose till no more dust washed through. Then it was set to drain on the kitchen table on some newspapers until the lag deposit could be examined closely. The material dries rather slowly, but I found that setting the screen on the floor over a heat vent helped a lot.

Home setup for sorting microfossils from other anthill material.

I found that the best technique was to scoop a little of the material out with a teaspoon and onto a sheet of plain paper, so it could be seen well. First, under good lighting, I'd look it over with a large, Sherlock-Holmes-style magnifying glass to cull out most of the small rocks from the material that needed further examination. I have a set of dental picks that I asked my dentist if I could have one time when they had gotten old and he was going to throw them away. They are good for fine prep work like scraping the dirt out from between the segments of a trilobite, and they came in handy here too. They have right-angle bends at the ends, sort of like a really skinny garden hoe. You just use them to scratch at the material, and whenever an interesting particle is found, it is shepherded off the white piece of paper and onto a green piece adjacent to it. Once you've been through the whole spoonful, just curl the white piece of paper in the middle and dump the leftover gravel into the wastebasket. Items of interest are then examined with a 20X monocular, which I find to be much more convenient than a microscope. You can see the fossils—which are in the 1 mm to 3 mm size range—plenty well, and your workspace isn't taken up with a bunch of equipment.

By far the most numerous fossils in my anthill material were ostracods. Ostracods are tiny (mostly freshwater) crustaceans, meaning that they are related to crabs, shrimp, and lobsters. The animal lives inside a miniscule bivalved shell, which is the part preserved as a fossil. Many of them look like miniature jelly beans or little grains of rice, but some have more ornate decorations on their shells. Altogether I found four different kinds. The jelly-bean type were the most numerous, but there were also a fair few of another kind that had the shape of a sprouting pinto bean and were covered with dark purple spots. I even found a few, very shiny ones with a sharp spire at the posterior end, and on which the suture between the two shells could be easily seen. Since the White River was deposited as stream and overbank material, these ostracods were probably washed from their watery homes by flooding, and deposited on the floodplain beside the river.

There were fossils from terrestrial creatures too. Less common than the ostracods, but present in fair numbers nonetheless, were bone fragments, tooth fragments, and eggshell fragments. Sometimes you could only tell that the bone fragments were bone by their texture; other times, by the shape, you could see that it had once been the end of a longbone—I have a nice once that shows the marrow cavity well. Tooth material is identifiable by the shininess of the enamel, and eggshell fragments by their thin, curved shape. Sometimes you can even make out the textured pattern on the convex surface of the eggshell.

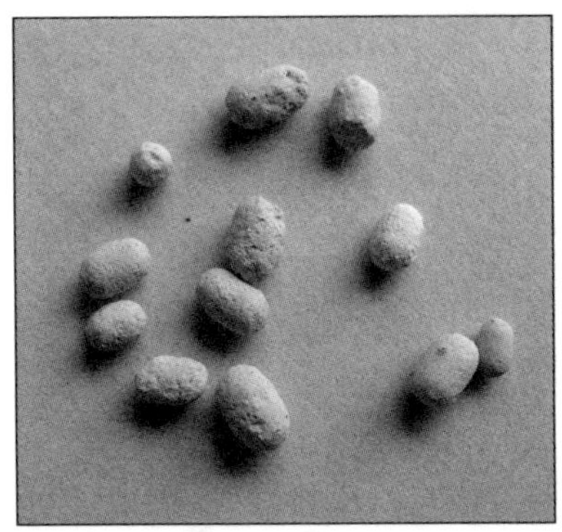

"Jelly bean" ostracods.

Eggshell fragments.

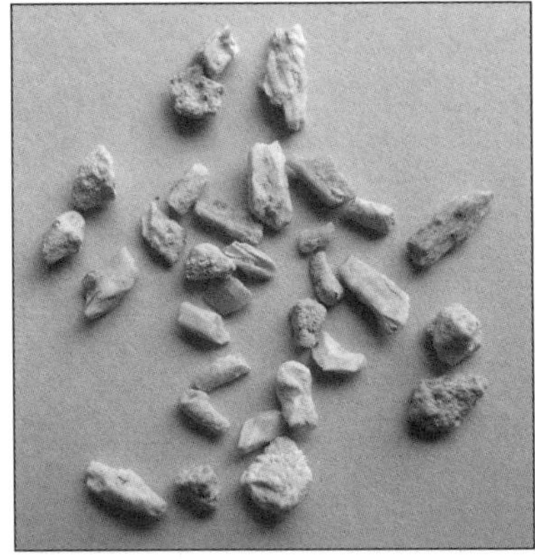

Bone and tooth fragments.

All of these latter fossils were too small and fragmentary to be identifiable as to what creatures they came from, but judging from their size, they would have originally belonged to animals perhaps the size of a mouse, and certainly no bigger than a squirrel. I don't know what the eggs were from—maybe birds nesting on the floodplain, maybe lizards. Before he passed away, a friend of mine, Karl Hirsch, became a self-taught world's expert on fossil eggs, and once he showed me how he coated fragments with gold and took scanning electron micrographs of them. Things like the thickness, microstructure of the shell, and pattern of lumps and pores on the surface could be used to identify the type of animal the egg came from, but I've never gotten into it that much, and you need the equipment they have at the university to do that kind of work anyway. For me, it was good enough to know they were fossil eggshells.

Folks who are really into microfossils carefully mount each one when they are done. They superglue it to the head of a pin and stick the pin into the cork of a test tube, then put the cork in the test tube to protect the fossil. They label the cork with an ID number and catalog each fossil separately. That's the way museums do it,

but it takes a lot of time and effort, and to me, one jelly-bean ostracod looks much like another; so I simply use different test tubes to
hold each type of ostracod, bone fragments, tooth fragments, eggshells, etc. with the fossils loose inside. This doesn't protect them as
well, but it's good enough for me, especially since these are not scientifically valuable specimens that need to be studied by scores of researchers.

It makes you feel a little bit like the Grasshopper, letting the
Ants do all the hard work for you and then just benefiting from their
labors, but it sure saves a lot of time! I found from a couple to a
dozen microfossils in each teaspoonful of anthill material—can you
imagine how long it would take you to collect those if they were spread
over several square yards?

4

Porcupine Cave

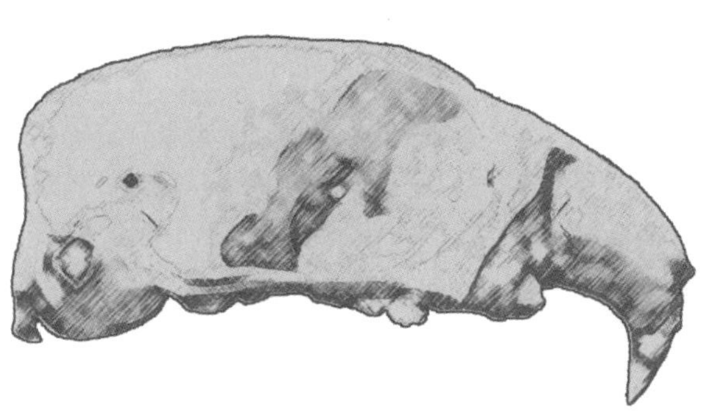

Fossils are found buried in the ground, right? Well, sometimes you have to go underground yourself to find them! During the summer of 1987, I participated in a dig jointly sponsored by WIPS and the Carnegie Museum of Natural History—excavating Porcupine Cave, high in the mountains of central Colorado near South Park. This natural cave was accidentally discovered by miners about a century ago when they broke into the cavern during a tunneling operation, but only in the mid-1980s was it recognized that fossils were to be found there. The Pleistocene sediments, as much as a

South Park. This high mountain valley is a graben or downdropped block that formed during a failed rifting episode that would have split North America in two. The Rio Grande river follows this rift valley further south.

million years old, contain the best-preserved high altitude fauna from this timeframe in the United States.

The site is on private land, and we were there with the permission of the rancher. The cave itself is around 9000 feet in elevation; we camped in the valley below, in a grassy meadow next to a small stream, in the company of resplendent wildflowers and unflappable cattle. You could make the 1500-foot climb and mile-and-a-half-long hike up to the cave if you wanted, or ride in a 4WD vehicle most of the way up. My trusty blue 74 Gremlin wasn't up to a steep climb on a rocky track, so I rode up to the cave along with some others in Dale's truck.

The entrance to the cave is the old mine tunnel, where you can walk almost upright as long as you are careful not to bump your head. I happened to have an old hardhat left over from some construction work at my last engineering job, which was absolutely invaluable. The tunnel isn't very long before you reach where it breaks through into the natural cave, and you have to climb down a long wooden ladder into the entrance chamber. On a ledge to your left as you descend the ladder is a modern representative of the reason many of the fossils are there in the first place: a packrat midden. Packrats have a well-known affinity for bringing interesting things they find back to their nests, and this includes bones, teeth, and other natural materials they find lying around on the surface. They had been doing this for millennia at Porcupine Cave, entering by a different, small natural entrance; over the years, vast layers of sediments containing the bones of Pleistocene animals had accumulated in various parts of the cave, obviously time-stratified. The objective of the dig was to collect representative samples from progressively older layers, and document how the fauna had changed over time.

Porcupine Cave is a dry cave, meaning that it lacks the ornate decorations found in scenic caves like Carlsbad. It is merely a series of rocky chambers floored by loose talus and a lot of dust. When I was there, the excavation was focusing on two of the largest chambers: the Velvet Room (named for a Velvet chewing-tobacco can that was found there—probably left by one of the miners), and The Pit. After descending the initial ladder, to get to the Velvet Room, you had to slither on your belly down along Tobacco Road, a low, wide tunnel that a rotund or claustrophobic person would want to avoid. But the Velvet Room itself is a roomy chamber a little bigger than my living room with a ceiling high enough to allow plenty of room to stand. Don had even strung a long, heavy-duty electrical cord in from a generator outside the cave's entrance, so that a homemade screening operation could be done right there, attached to a vacuum cleaner to suck in the dust.

The screening apparatus consisted of a stacked series of wooden frames floored with differing mesh sizes of hardware cloth and screen. The largest mesh was at the top, where you put the sediment in, and caught the bigger rocks and particles, mostly lumps of clay. Each successive screen below had a smaller mesh size, and retained the next-largest particles, including some of the fossils. Jaws would be caught early on, whereas tiny teeth would go through most of the screens and be caught by the last one, below which the dust was slurped into the shop vac. By screening right there in the cave, it minimized the work we had to do in carting material out of the cave, especially slithering up Tobacco Road.

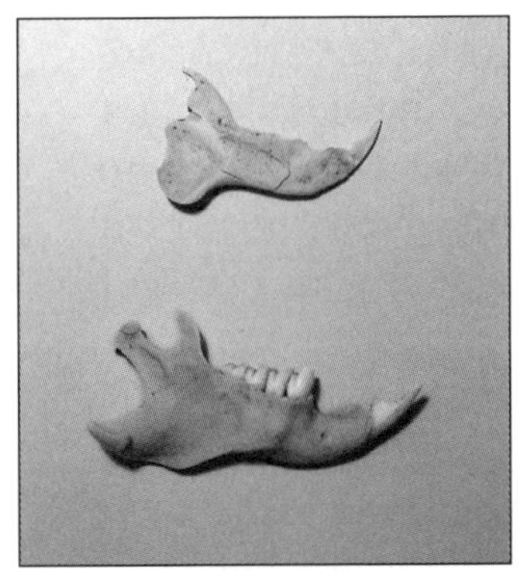

Lower jaws of squirrel (top) and prairie dog (bottom).

All of the fossils in the cave were small, because they were things a packrat could carry. The vast majority were from small animals, although sometimes a tooth or toe bone of a larger animal was found. Since the mission was to identify how the fauna had changed over time, the dig was run much like an archaeological excavation, with quadrants gridded off and layers only an inch or so thick scraped off and kept separate from one another. Each batch of fossils was carefully tagged with an ID number that located it both geographically and stratigraphically within the excavation.

I worked in the Velvet Room the first day I was there. You'd sit next to your assigned grid spot, and carefully scrape the dirt into a bag with a flat-sided trowel. When you filled your bag, you'd take it over to the person running the vacuum-screens, who would sift it through and then pick the fossils out from the rocks and dirt lumps on each screen, and then individually wrap and/or bag the fossils. There were teeth and jaws and other bones from mice, voles, rabbits, and even black-footed ferrets. There were also occasional bird bones, coyote teeth, snake vertebrae, and a later team even found a camel toe bone, I heard. Tony Barnosky and Don Rasmussen, the dig leaders, explained to us how to discern a rabbit tooth from a rodent tooth and such. Periodically someone would take bags of material out of the cave and set them near the entrance, where another group was screening outside (it got pretty dusty, because there was no vacuum, but the wind blew most of it away—after making sure your eyes got their fair share).

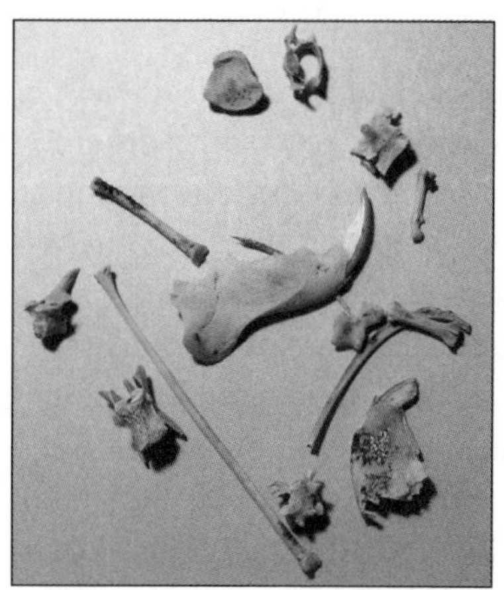

The bones were small things that a packrat could carry in.

The second day, I worked in The Pit. To get there, instead of going down Tobacco Road, you made a left into a long passage where you could walk, stooped over. Then you came to another wooden ladder, down which you climbed into The Pit. The sediments there were stratified into a series of alternating layers of hard, grainy stuff and fine dust. Tony thought these might represent the comings and goings of glacial and interglacial spells, with the dust representing the colder, drier conditions and the pea-sized grains having been formed when a warmer, moister interglacial period arrived. Again, it was excavated archaeology-style according to a grid, but The Pit was too far from the entrance to be set up with a vacuum-equipped screening apparatus, and you can't screen without one inside the cave—it just makes too much dust. So they had a whole slew of canvas money bags (the kind you see bank robbers make off with in the cartoons) that we filled with sediment and marked as to horizon. At lunchtime and at the end of the day, the whole bunch of us would form a long line between The Pit and the entrance and hand off the bags one to another until they were all amassed at the entrance of the cave. Then we'd form another such line between the cave entrance and where the trucks were parked, and thusly move the bags of "treasure" into the back of one of the pickups to be transported back to camp.

Not everyone who participated in the excavation wanted to work inside the cave, and indeed, there wouldn't have been room for them all if they had. Some people preferred to work the screens outside the cave entrance, and another set stayed at camp and sorted through the material from The Pit. Sediments from the grainy layers had to be soaked in the creek overnight to soften them, before they could be dried and picked through for fossils. The creek was also a great amenity to have for washing yourself in after a long day of work in the cave—a change of clothes and a wash of face and hair made a person feel really refreshed after grubbing about in the dust all day.

I only went to Porcupine Cave that one summer, but sixteen years later, the work continues. Tony left the Carnegie and the project was picked up by the Denver Museum of Natural History (recently renamed the Denver Museum of Nature and Science), which still operates annual excavations there for much of the month of July. In the years since I was there, several other caches of bones have been found in the cave, which are thought to have fallen in through natural traps. Some really exciting finds have been made, including those of the world's oldest mountain goat, coyote, black-footed ferret, and

several teeth from a cheetah. During the winters, the bones collected each summer are "picked," identified, and catalogued into the
Denver Museum's collections, and a major monograph on this fascinating Pleistocene cave fauna is currently in the works.

5

Hungry Hollow

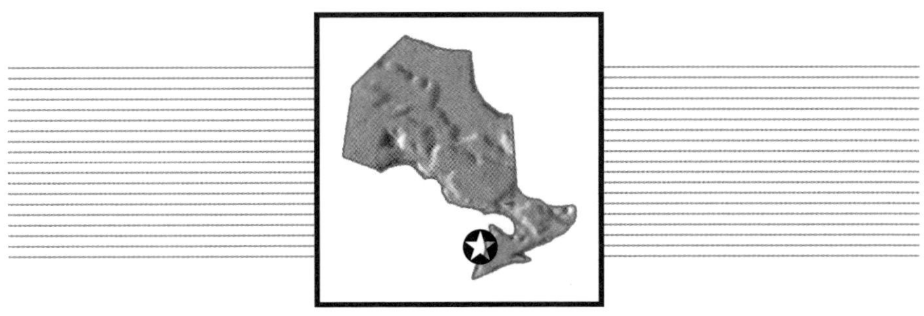

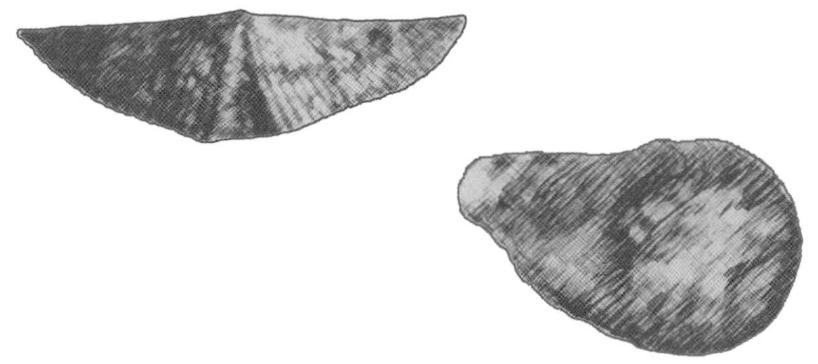

In the summer of 1988, I went back to Michigan to visit my parents and help them sort through 26 years of accumulated household goods in preparation for my dad's retirement and their planned move to Colorado. While I was there, we decided to take a break from the drudgery and make a fossil-hunting foray into nearby Ontario. My sister had taken a geology class at the local community college and received a map showing how to reach a location called Hungry Hollow, near Arkona, where many invertebrate fossils could be found, and my mom and my aunt had collected some nice horn

My Aunt Dotli hunting fossils at Hungry Hollow.

corals there about ten years before.

My mom is the one who encouraged me in my love of fossils. My dad is only mildly interested, but is always a good sport and comes along. We loaded up the truck with collecting paraphernalia one bright, hot day, crossed the bridge at Port Huron into Sarnia, and began to follow the map.

We took a few wrong turns, and had to do a little driving around before we finally came to the Ausable River and the spot on the map. Here the river has cut a gorge through Devonian marine strata of the Widder Beds, Hungry Hollow Formation, and Arkona Shale. The Hungry Hollow Formation is a grey limestone and the most productive unit for fossils. The sides of the gorge can be steep, and you do most of your collecting by sorting through the chunks of rock that have weathered out of the slope and lie on the flat near the bank of the river. If you are really ambitious, you can chip away at the limestone with a rock hammer, but we found it to be much easier just to look for specimens that had already weathered out. There are so many of them that it's really not worth working for the ones remaining in the cliff.

Unidentified globular brachiopod (top) and *Mucrospirifer* (bottom).

Mucrospirifer as more commonly found.

Horn corals (*Zaphrentis prolifica*) are by far the most common. They are typically a couple of inches long and quite well-preserved, showing the details of internal septae on the big end of the "horn." Brachiopods (*Mucrospirifer arkonensis*) also weather out of the Widder Beds above and tumble down to the river. These are an elongate, winglike form, and are usually missing the "wingtips." From time to time you'll also find some plumper, more globular brachiopods, but those are less common than the *Mucrospirifer*. Crinoid columnals also occasionally appear, but make up a minor element of the fauna.

Zaphrentis horn coral showing internal septae.

We only spent a couple of hours collecting that day. For one thing, the heat brought on one of my killer migraines. All I had with me was some aspirin, which is pretty impotent against a migraine, and I tried to splash cold water from the river on my face to abort the headache, but to no avail. I had to stop hunting, go lie down in the air-conditioned car, and put a ziploc full of ice on my forehead. My mom continued for a while longer, but in such prolific strata, it doesn't take long (even if you are selective) to accumulate the 50 lbs. of invertebrate fossils allowed to a non-citizen removing them from Canada for non-commercial purposes. With boxes full of horn corals and a lesser quantity of brachiopods, we climbed into the truck and retraced our route back to Michigan. It was a nice little break, but a lot of packing and painting still awaited us.

6

Tracking Dinosaurs along the Western Interior Seaway

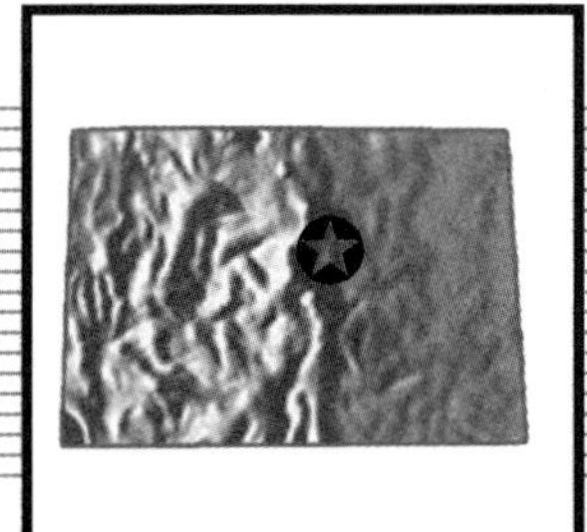

S ometimes, collecting fossils is the worst thing you can do. Say what? Well, when it comes to dinosaur footprints, most of the time they are better left in place, and protected for others to view and enjoy.

About 20 miles south of where I live is the tiny town of Morrison, Colorado. This quaint little berg has the distinction of being the source of the moniker of the famous Jurassic Morrison Formation, which is exposed widely in the western states and produces a fabulous array of dinosaur bones, including those at Dinosaur National Monument. For many miles along the eastern fringe of the foothills—where the plains of eastern Colorado give way to the Rocky Mountains—runs the Dakota Hogback. This is a steep, linear ridge

Dinosaur Ridge. The roadcut excavated when Alameda Road was put through exposed the Dakota Formation sediments.

or cuesta that is faced on the east with the erosion-resistant Dakota Sandstone beds that protect the softer Morrison strata that lie beneath them, and which are exposed on the west side of the ridge. This hogback, now known locally as Dinosaur Ridge, is the site of the first significant dinosaur bone discoveries of the late 1870s—which ignited the infamous Cope and Marsh "bone wars."

The ancient beach sediments were tilted at about a 40° angle when the Rocky Mountains were uplifted.

But the eastern face of the ridge is what interests us here. The Cretaceous Dakota Sandstone was laid down as beach sands along the shore of the Western Interior Seaway. During much of the Cretaceous, seafloor spreading rates were high, and hot rock occupies a greater volume than cold rock. This increased volume of rock near the oceanic spreading ridges, along with the absence of polar ice caps, was enough to raise sea levels worldwide, and create a number of epicontinental seas—areas of continental crust that were flooded by shallow waterways. One of these, the Western Interior Seaway, cut North America in two and connected the Gulf of Mexico with the Arctic Ocean. It is sediments laid down in this seaway that form the thick chalk deposits of Kansas that produce such fantastic fossil fish.

The present Rocky Mountains had yet to rise, but the Ancestral Rockies, which towered over central and western Colorado 300 million years ago, still commanded enough relief to keep that part of the state above water. So the shoreline of the seaway paralleled the present Front Range, and along its beaches strolled vast herds of dinosaurs, leaving their footprints in the moist sand. It has been estimated that this formation in Colorado and New Mexico alone preserves mil-

Rooter and dinosaur track.

lions of dinosaur tracks, earning it the nickname "the Dinosaur Freeway."

A couple of times while I was in grad school, our professors took us down to Dinosaur Ridge to view the footprints and other sedimentary structures. As well as tracks, different horizons preserve trace fossils and ripple marks in the sandstone. During the 1980s, you could walk right up to the tracks. I have a great photo of Rooter sitting next to an iguanodontid (*Caririchnium*) footprint. But look at the other photo, and notice the circular ring around the track: that's the work of vandals trying to chisel out the dinosaur footprint and steal it, but they gave up. A few tracks were actually stolen, and now there are only potholes left. First a chain-link fence was installed, and then in 1990 a group called Friends of Dinosaur Ridge was formed to push for protection of the area, which is now a National Natural Landmark. There is a nice visitors' center, and on "Open Ridge Days," volunteer guides station themselves along the road to interpret the footprints and sedimentary structures for the public. This is a lot of fun, and for a couple of years in the mid-1990s, I served as one of these guides.

The ring around this track is the work of vandals trying to steal the footprint.

Most of the tracks in the Dakota Sandstone, a Middle Cretaceous unit around 100 million years old, are of tridactyl (three-toed) ornithopod dinosaurs, prob-

Positive-relief theropod tracks under a cliff near Morrison.

ably *Iguanodon* or a related species. Before the discovery of footprints, the posture of *Iguanodon* was highly speculative. Benjamin Waterhouse Hawkins reconstructed it as a gigantic, sluggish-looking, belly-dragging quadruped for London's Crystal Palace exhibition in 1854. You can still visit his reconstructions in Crystal Palace Park, Brixton today. Later in the 19th century, American paleontologists restored *Iguanodon* as a kangaroo-like biped. But tracks from the Dinosaur Freeway show that it was what is called a facultative biped, that is, it walked on all fours much of the time, but could rear up and walk on just its hind legs when it wanted to, perhaps for a speedy getaway. This is shown by the presence of manus (hand) prints in the same trackways as pes (foot) prints. Indeed, the average person may not find dinosaur tracks as exciting as a big, mounted skeleton, but they have much to tell us about the behavior of these fascinating creatures.

The sheer numbers of these iguanodontid footprints indicate that these were probably herding animals. Most of the trackways are oriented generally north-south in the Dinosaur Ridge area, presumably parallel to the shoreline at that time, and it is currently thought that they may represent the seasonal migrations of herds of these dinosaurs.

But ornithopods are not the only dinosaurs who left their tracks here. Ornithopod tracks, while tridactyl, have relatively short, fat toe impressions, and frequently also show evidence of a heel. Theropods—carnivorous dinosaurs—also inhabited the area, and occasionally also left their tracks. These also are three-toed, but the toes are much longer and more slender, and do not show a heel. They quite resemble the tracks of overgrown birds, and the first theropod tracks discovered early in the 19th century in New England were in fact thought for some time to be bird tracks. It is now known that they are dinosaur footprints, but the initial assessment was not too far off, since today most paleontologists believe that birds are the direct descendants of small, meat-eating dinosaurs.

Most of the footprints at Dinosaur Ridge are impressions; that is, they look like footprints—the shape of the foot is indented into the rock. But closer to Morrison, just east of town, is a set of theropod tracks high up underneath a cliff. These are preserved in positive relief—the footprint bumps out from the rock. They were formed

Plaster cast of varanid lizard tracks.

when the dinosaur made its tracks in a muddy area, and they were later filled with sand. Over millions of years, the mud turned to shale and the sand to sandstone. Since sandstone is more resistant to erosion than shale, the shale weathered away, and you are looking up at the underside of the tracks!

So even though these footprints are protected and a person can't collect them, there is a way to have a footprint in your personal collection: by making a cast. I've never done this myself, but one of my grad school classmates, Kelly Conrad, has. Latex rubber is painted onto the track, and after it dries, it is filled with plaster of Paris. This gives you a positive-relief reproduction of the track that you can keep to study or just enjoy, without destroying the original. Kelly gave me a nice cast of a couple of varanid lizard tracks to keep, and I use it as part of my school-talk kit—all without disturbing a scientifically valuable resource.

7

Prospecting "Down Under"

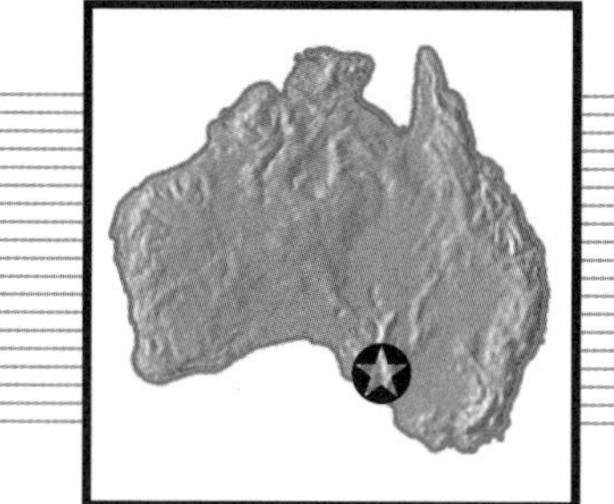

In May of 1989, I got the chance to visit the place that I'd always said was my first choice of a faraway destination: Australia. Fortuitous circumstances had placed two pairs of my friends down under that year. Nancy and Alex were living in Sydney while he worked on an overseas assignment for his company. Jordan and Ruth were spending the year in Adelaide on a teachers' exchange program. So I figured it was the perfect time for me to visit the southern continent.

Looking south towards Maslin Beach.

After a short stopover in New Zealand, where I explored Auckland and the hot springs at Rotorua, I landed in Sydney on a crystal blue morning. I was greeted by Nancy, who told me that it had been raining for three months straight until that very day. (It's nice to live in Colorado and take its sunny weather with me when I travel; the same thing happened three years later when I visited London.) I spent a week in the Sydney area exploring such places as the Australian Museum, which has a very nice set of dioramas on prehistoric Australia, and the Royal Botanic Gardens, with its extensive plantings of native and exotic cycads. My interest in cycads goes back to my readings about Mesozoic landscapes when I was a kid— cycads are always pictured alongside the dinosaurs, because their

Looking back at Hallett Cove from Maslin Beach.

heyday was during those times. I have been growing the living cycads as houseplants since I got my first apartment, and have been a member of the Cycad Society since its founding in 1977. I got my first chance to see cycads growing in their native habitat when we visited Kur-in-gai National Park about an hour's drive north of the city. Beautiful *Macrozamias* dotted the understory of an open eucalyptus forest. Cycads are gymnosperms (non-flowering plants)—more closely related to pine trees than to the ferns or palms they superficially resemble—and some of the male plants even had mature cones on them. I also toured the reptile park at Gosford, which houses a nice variety of the native goannas (varanid lizards), other reptiles, and two of only ten duck-billed platypuses in captivity, named Eb and Flo. Then it was on to Adelaide to visit Jordan and Ruth.

Jordan trying to net a fossil sea urchin out of the cliff with a golf ball retriever
(well, it worked the first time).

Jordan is one of the founders of WIPS and has been fossil-hunting in the States for over three decades, so of course, when he got to Australia, he had to check out the rocks. By the time I arrived, he knew all about the local geology and paleontology. He took me to the South Australian Museum—not as big as the one in Sydney, but quite interesting nevertheless. It has three huge skeletons of (modern) whales hanging from the ceiling in the lobby. We also visited the botanical gardens in Adelaide and saw more cycads. I shared my knowledge of these fascinating plants with him, just as he had shared his expertise in invertebrate paleontology with me when we first met. He also took me to the University of Adelaide geology department, where he was teaching, and showed me their nice collection of Ediacaran fossils. But what we both really wanted to do was spend some time outdoors collecting our own fossil specimens.

Adelaide sits in a jewel-like setting on the west coast of South Australia's bulge, at the east end of the Great Australian Bight. To the north are Lake Eyre and the parched outback basins; to the east, fertile wine country and the eucalyptus-cloaked hills that divide the Murray River drainage. Jordan and I set out to hunt fossils on one of those bright, clear winter days when the Southern Ocean mirrors the turquoise sky and the beach begs to be experienced barefoot.

Fossil echinoid (urchin) in limestone
(yes, we got this one out).

Driving south from Adelaide, we stopped at Hallett Cove. Here, the bedrock is sculpted into monolithic forms which rise out of the sands like beached whales. And, like marine mammals carrying the scars of their encounters with propellers, these massifs, too, bear striations on their backs; these not from human intervention, but as testimony to the glaciers which crawled across the southern hemisphere during the Permian. Boulders frozen in the base of the ice sheets ground grooves in the bedrock as they bulldozed a path northward from Antarctica (then joined to Australia).

We continued south along the shore to Maslin Beach. Here, vermillion cliffs bisected by the Hallett Cove Sandstone flank a narrow beach that is licked by constant breakers. The Hallett Cove conformably overlies the limestone of the Eocene-Oligocene Port Willunga Formation and the clays and sands of the Late Eocene Blanche Point Formation. Near where we parked, a lone eucalyptus tree stood as a sentinel overlooking the ocean. Streamers of ice plant,

Typical echinoids and detached spine.

a mesembryanthemum or "living rock" plant, cascaded down the hillside, covered with pink and white aster-like flowers. Here the sediments of the Blanche Point look soft enough to crumble in your hand, and tempt you to reach up and pluck out the globular echinoid fossils that pepper the cliff face... and in a few places, you can. But when we tried to use a golf ball retriever to net some of the fossil urchins just out of reach, we discovered that in other places the rock is cemented like, well, cement—with the same calcite that has filled the interiors of the shells with crystals. The golf ball retriever was useless, and we had to break out our rock hammers.

Thus armed, the ancient sea floor yielded its entombed inhabitants to us. Our collecting was done in the Tortachilla Member of the Blanche Point, a sandy-ferruginous limestone near its base. Common are the heart urchins, and the sediment is littered with millions of their detached spines. We also collected some clams. The fauna has been interpreted as representative of a shallow-water marine benthic (bottom) community; during the Eocene, this area was part of the St. Vincent Basin on the

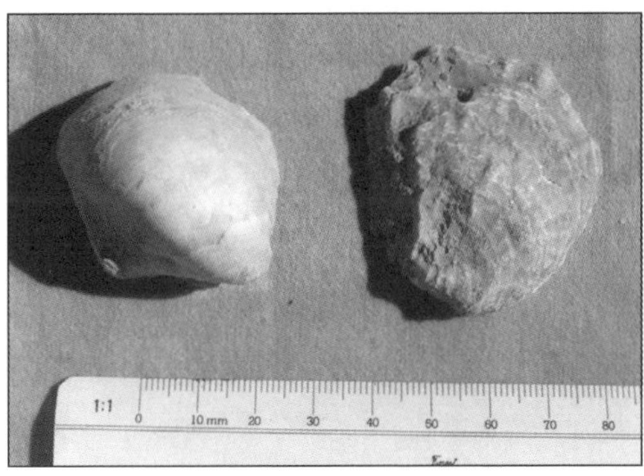

Fossil bivalves from Maslin Beach.

passive margin of the Australian plate, and is thought to have had limited communication with the open ocean.

It took us about two hours to collect all the fossils that it is practical to stuff into your luggage on an international flight, where baggage limitations force paleontological treasures to compete with the obligatory presents for the in-laws. Then we climbed the hill back to the car and went off to hunt kangaroos (with a camera, of course).

One of the best echinoids I collected. Note the fine detail.

8

Douglas Pass

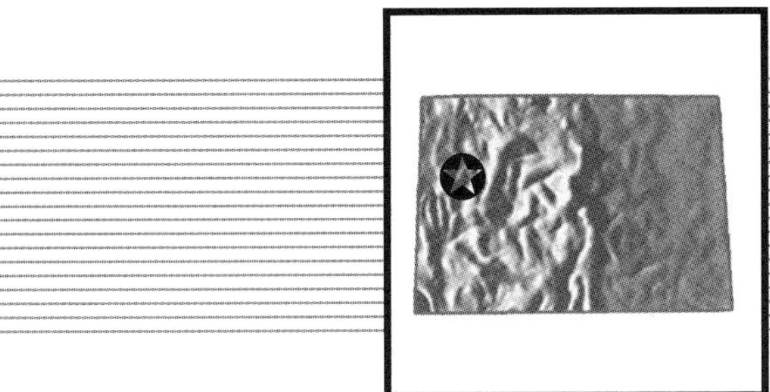

I've been to Douglas Pass several times over the years, always as a side-trip from another dig. I was first introduced to this world-renowned collecting locality in 1986, when I was part of George's team digging for small vertebrates at Fruita. Every few days, to break the monotony, George would give us a "day off" and we'd do something different. One time we loaded into the vans and headed northwest to Douglas Pass.

The valley of the Colorado River, where it emerges from the Rocky Mountains and heads southwest into the canyon country of

View north from Douglas Pass.

Utah, is ringed with higher ground on three sides. Grand Junction, the largest city on the Western Slope, sits in the middle of the valley. To the east looms the dark and brooding Grand Mesa, a high volcanic plateau. The southern rim of the valley exposes Mesozoic sediments dipping to the north, and it was in these that George's dig was conducted. The north rim of the valley is lined with what the locals call the Book Cliffs, which are various Tertiary exposures. To the west lies open desert, through which the river, fringed with green tamarisks and cottonwoods, flows like a lifeline.

You get to Douglas Pass by going north on highway 139 until it climbs and crests the Book Cliffs. Exposures at the top of the pass are of the Eocene Green River Formation, which has its origins in sediments laid down in three large lakes that formerly covered much of the area near the juncture of Wyoming, Utah, and Colorado. Different members of the formation produce a variety of fossils—leaves, insects, fish, even other small vertebrates occasionally. I have collected leaves in the Green River in Utah and fish in Wyoming, but at Douglas Pass, the sediments are best at yielding insects.

Camping at Douglas Pass.

Green River shales can be very fine-grained, which is why they preserve fossils in such exquisite detail. The bottom waters of these lakes were frequently anoxic (low or lacking in oxygen), which excluded any animals that might have scavenged or stirred up what settled to the bottom. This allowed delicate organisms to remain intact long enough to become buried by more sediment, the first step in fossilization.

Roadside rock exposure at Douglas Pass.

One convenient thing about the Douglas Pass exposures is that you can drive right up to them, so it's a suitable collecting spot even for those who aren't up to long treks through the desert. After a series of very long switchbacks, the highway comes to the crest of the pass. Right there, on the east side of the road, is a small, unmarked gravel road. You can either park there and collect in the exposure of the roadcut, or go further up the gravel road and collect in any one of the numerous exposures along the way.

If you want to do any serious digging, you'll need some tools like a rock hammer and chisel, but all of these exposures have a lot of loose talus at the bottom, and I've found that it's easier—and quite as productive—just to sit near the bottom and examine the loose pieces that have weathered out. The shale here splits into pieces usually no more than a quarter of an inch thick, and a few inches square. Just get comfortable and start examining both surfaces of all the loose pieces within your reach. You'll find a lot of little tubular fragments of plant stems, and fairly often a nice little

Typical Douglas Pass material: plant stems and an insect larva at upper right.

carbonized impression of an insect larva that somewhat resembles a modern grub, segments and all. I'm no bug expert, so I've never made any real effort to identify these specimens taxonomically, but the first time I was there I found a perfectly-preserved, complete small insect of some kind, which looks to be a gnat or a small fly. All

My prize Douglas Pass specimen: a 3 mm long gnat or small fly.

I can really say is that it is in the Order Diptera and looks much like a modern black fly, and is also the right size—around 3 mm long. It's really an amazing specimen, showing the legs and even one eye and wing if you look closely. I'm glad I found it seventeen years ago, because these days I need to put on my strong glasses even to see it at all! A couple of other guys I know, Mike and Bill, have extensive collections of Douglas Pass material and have even compiled a large catalogue of different insects that can be found there.

You might say, what's so exciting about fossilized insects and broken plant stems? Well, it depends on your perspective. A tiny fly may not impress as many people as a big dinosaur bone, but it's a lot easier to amass a collection representative of the ancient ecosystem at Douglas Pass than at most other sites. And like I mentioned before, the site is on public land and easily accessible even to creaky old ladies (like my mom) who have the spirit for collecting fossils, but not the ability to trek in miles to a site. That's why we stopped there a second time, in 1990, when Allison was a baby and my parents and I were driving up to Wyoming, pulling their camper. At a parking area near a transmission tower, there was no traffic, and my dad could watch Allison while my mom and I hunted for fossil treasures—something that would have been impossible at most collecting sites.

One thing you do have to be careful of with Douglas Pass specimens is their delicacy. The shale here is quite soft, and the fossils are preserved as thin films of carbon on the surface of the rock. So whenever you find a keeper, you have to wrap it carefully in toilet paper and place it in a protective container (I find that Pringle's potato chip cans work great for holding a lot of small specimens). And don't put these fossils in your garden as decorations when you get home, like you might do with a nice piece of petrified wood—a little rain and your fossil will be gone, washed right off the rock. But if you keep them safe and dry at home, like in a drawer or display case, they can really be beautiful. Just like the scenery at the pass—dense spruce and aspen forests, typical midaltitude Colorado; and panoramic vistas of the river valley far below.

Cañon Pintado pictographs, Kokopelli in center.

One last thing that you'll want to make sure you do if you make a trip to Douglas Pass is stop at Rabbit Valley if you're coming or going from the south, or the Cañon Pintado pictographs if your route is down highway 139 from the north. Rabbit Valley is actually an exit from I-70 just this side of the Utah border, but close to where the Douglas Pass road takes off. It preserves the Jurassic Morrison Formation and some of the large dinosaur bones it contains, like those of *Camarasaurus*. It's a small park with an interpretive trail where you can see bones still embedded in the rock and learn about the geology of the area. Cañon Pintado is a stretch of the valley through which highway 139 winds that cuts through the Lower Jurassic Entrada Sandstone and displays artwork of Native Americans who once inhabited the area. The pictographs are the work of the Fremont culture and date to around a thousand years ago. One of the figures has been identified as Kokopelli, the hump-backed flute player who plays a prominent role in southwestern rock art. The first description of these pictographs by a European was written by Fray Escalante in 1776, indicating that the Dominguez-Escalante expedition to find a better route to California from Santa Fe passed by this area.

9

Fish of Fossil Lake

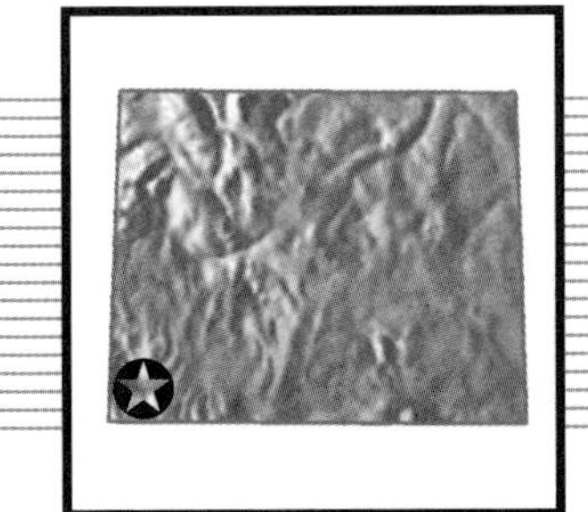

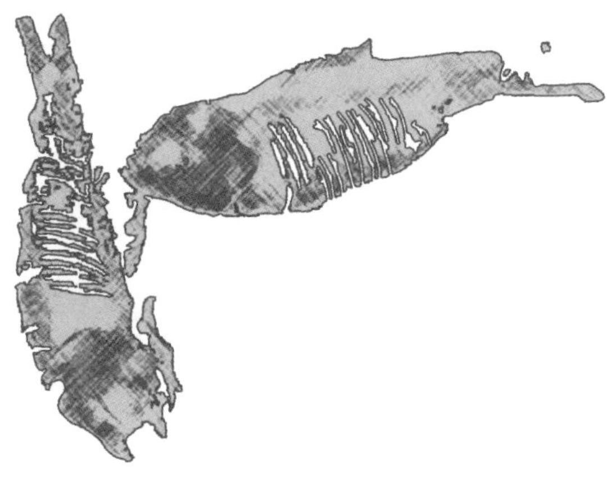

After leaving Douglas Pass and the Cañon Pintado pictographs in August of 1990, my parents, Allison, and I continued north into the far southwestern corner of Wyoming. In this area, the Green River Formation outcrops are those remnants of sediments laid down in ancient Fossil Lake. Here are preserved some of the best fossil fishes in the world, and in such quantities that the herring *Knightia* has been cited as the most commonly-collected vertebrate

Camping at the Warfield Quarry.

fossil in the world, with over 20,000 specimens having been unearthed in 1978 alone. It is thought that *Knightia* was a schooling fish that was sometimes subject to mass mortality due to low oxygen conditions in the lake, resulting in these amazing concentrations of fossil fishes. Because of this abundance, *Knightia* has been honored as the state fossil of Wyoming.

The quarry headquarters.

This prehistoric lake system—which consisted of three lakes: Uinta, Fossil, and Gosiute—is one of the largest and longest-lasting in the fossil record. Although the lakes varied in depth and areal surface extent over time, the system as a whole persisted for some 20 million years, forming around 60 million years ago and disappearing near the dusk of the Eocene. Each of the three lakes represents a distinct faunal facies in a subtropical or warm-temperate climate, with most of the best fish coming from Fossil Lake, which was the smallest, deepest, and shortest-lived of the three.

Most of you who have been to this part of the country will probably be familiar with these fish fossils as those showcased at Fossil Butte National Monument near Kemmerer. *Knightia* specimens are also commonly sold in souvenir shops across the state. But what about collecting your own specimens?

Because most of the best outcrops are either on privately-owned land or BLM land with the mineral rights leased, you will have dubious luck in just meandering about the countryside looking for specimens. Fortunately, for a small fee ($20/day or so), you can dig for your own fish to keep at one of the several commercial quarries in the area; we chose Warfield.

Even with directions, the drive takes you quite a way off the beaten path on poorly-maintained gravel roads, and (of course) it was getting dark by the time we were pulling in. Nevertheless, we made it to the parking area with my parents' camper, and it was nice not to have to set up a tent, especially with the baby along.

The weather was beautiful the next morning, and as soon as one of the guys from the quarry had opened the shack, we paid our fee and my mom and I headed over the hill and into the fish quarry, which is only a short walk from the parking lot. My dad had volunteered to watch Allison so that mom and I could dig, which we really appreciated. At five months old, she wasn't crawling yet, so he just needed to stay in the camper and keep her entertained between naps.

I'm holding Allison with our dogs in the background.

The quarry was a long trench maybe 6 feet deep, and you just picked a likely-looking spot and started digging. The shale here is harder and more consolidated than at Douglas Pass, and you really do need a rock hammer to be able to dig out the fossils. It's also more of a light beige color, in contrast to the greyish color of the Douglas Pass sediments. There was a fair breeze, which was good to keep us from becoming too hot out in the baking August sun, but it kept blowing dust in our faces, which was annoying.

The Green River is highly variable in its yields of fossil fish. Some layers preserve mass mortalities that can reveal hundreds of complete fish in a square meter of rock. Unfortunately, either we were unlucky, or the quarry owners save the best exposures for their com-

A pair of nice *Knightia* specimens that I bought later at a rock show. Would that our finds had been so good!

mercial operations, because all of the fish we found were widely-scattered and fragmentary. I wouldn't have even known they were fish fossils had I not known that was what was found there—just a few ribs or small vertebrae at a time were all we uncovered all day. Most of them weren't even worth keeping. We had a good time, but it was mostly for the fun of the hunt and not because of any of our finds. I think you'd want to spend a whole week there if you really wanted to come back with any nice specimens. The only Green River fish I have in my display cases are some that I bought. For what you pay for a day of digging, you can easily buy a nice, complete fish or two at a rock shop or show. On the other hand, then you don't have the same appreciation of the context they come from.

I wouldn't drive halfway across the country just to dig at one of these quarries, but if you happen to be in the area anyway, or passing by on I-80, it certainly makes a fun detour. It got me out of the house and gave me a break from the constancy of caring for a new baby, so I didn't really mind that we didn't uncover any good fish. Where else can you do that?

10

Hunting for Fossil Cycads

in the Petrified Forest

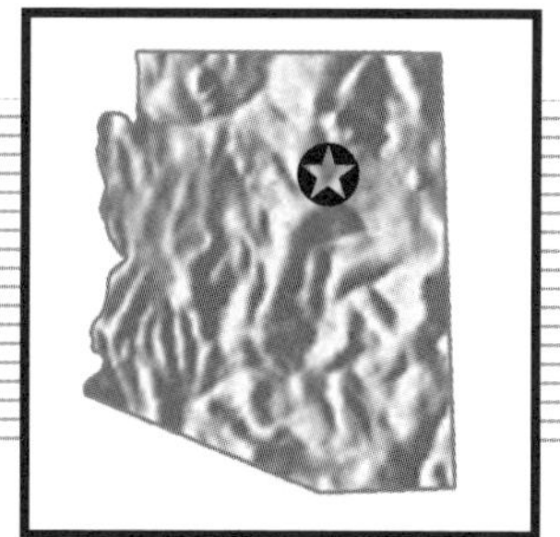

As part of a cooperative venture between the Western Interior Paleontological Society (WIPS) and the Petrified Forest National Park, I had the opportunity to do some fieldwork in the park in October 1992. The rocks exposed there are mostly the Upper Triassic Chinle Formation, a series of interbedded sandstones, claystones, and shales representing lakebed and floodplain deposits. All of our work was to be in the Lower Petrified Forest Member of the Chinle, in rocks approximately 220 million years old.

Petrified Forest panorama.

At the time these sediments were deposited, this part of Arizona was a low floodplain environment near the southern shore of a seaway that reached across Nevada and communicated with the Pacific. High, volcanic mountains to the south provided ample sediment in the form of volcanic ash, which was redistributed by a network of streams that dissected the lowlands on their way to the sea. The climate was generally arid, but strong monsoonal rains delivered plenty of moisture at certain times of the year, supporting the vast forests of massive conifers that give the park its name.

Malcolm collecting sediment at the Dying Grounds. We took it back to the condo to screenwash for microvertebrates.

Five of us drove down from Denver to meet with park paleontologist Vince Santucci and assist him with various research and reconnaissance projects. We were joined by a sixth WIPS member from Arizona, and all shared a condominium generously provided by the park service for our use.

Although the park is most famous for its amazing deposits of petrified wood, especially the Arizona state fossil, *Araucarioxylon arizonicum*, it also contains fossils of a remarkably varied vertebrate fauna, including remains of large amphibians, many types of thecodont reptiles, a variety of fish, and some of the earliest dinosaurs. Leaf fossils are found in certain claystone layers and represent cycadeoids, ginkgos, ferns, horsetails, and conifers; the stems of two types of true cycads (*Lyssoxylon* and *Charmorgia*) have also been found in the park.

Phytosaur scutes (above) and palatal tooth plate of lungfish (below, center) from the Dying Grounds.

The first day we were there, Vince gave us a tour of the park to acquaint us with the geology and paleontology there. He described the different rock layers and took us out to a place called the Dying Grounds, which is in between the Teepee Buttes and Blue Mesa, in the backcountry of the park. Here the fine shale, which weathers to the "popcorn" texture typical of bentonite, preserves the remnants of shallow lakes and streams. We found phytosaur scutes and a nice lungfish tooth plate, and gathered some bags of material to take back to the condo for screenwashing—hoping that evidence of some microvertebrates might be found. The next day we split up into teams and set off on various forays.

I was excited about the possibility of discovering some cycad fossils since I grow many of the modern genera as a hobby. So while part of our field crew went to survey the Blue Mesa area for evidence of large vertebrates, Doug and I chose to work in another area and look in one of the grey claystone layers for plant fossils. (I cannot reveal locality information due to problems with vandalism in the park. A recent survey found that an average of 3-1/2 lbs. of petrified wood per vehicle leaves the park illegally. Ever heard of Fossil Cycad National Monument in South Dakota? It isn't even a National Monument any more because vandalism has denuded it of all the cycads.)

I traced the grey claystone layer around the base of a small hill, looking for a place where it would be relatively easy to remove the overburden and get to some fresh rock. A deep, narrow arroyo on the west side of the hill provided me with the access I needed. It was late in the afternoon and the arroyo was in shadow by the time I began quarrying, but in the hour or so that I worked before my rendezvous with Doug, I uncovered enough plant hash (mashed up leaves and stems) to determine that I had found a promising locality.

Doug, Vince, and I returned the next day with a large pick for removing the overburden in an expedient manner. Once this had

Zamites powelli frond from Cycad Wash.

been done, Doug and I remained to work the quarry. It's hard to explain, but I could just "smell" cycad fossils (then again, maybe it was merely optimism). But before lunch I began finding detached leaflets that looked very much like those of a modern *Zamia* in their shape and venation. Doug was working on the opposite side of the wash and stratigraphically a little higher than I was, where he was finding mostly plant hash, some of which looked like the giant horse-

Zamites fossils and modern *Zamia* leaf.

tail *Neocalamites*. It was after lunch that, getting tired, I was swinging my rock hammer rather nonchalantly at some broken, blocky stones when out popped a beautiful pinnate leaf! A spontaneous "Hallelujah! There it is!" sprang from my lips. The frond was not complete, but the rachis and about a dozen pinnae on each side were preserved as a black carbonaceous film in the rock. Doug was suitably impressed and immediately stopped what he was doing to help me search for the counterpart, which he found on the floor of the arroyo. It lacked the carbonized film but showed the perfect mirror image as a shallow impression. It was spooky to see how closely the frond resembled one of a modern *Zamia*, given the incredible timespan that separates them. The fossil lacked the conspicuous constriction of the pinna bases common to the modern *Zamias*, and the pinnae were squared off at their ends rather than tapering as in the modern forms in my collection. But it was obviously a cycad-like leaf.

The adrenalin began flowing and I didn't feel tired any more. I just knew there were more cycads waiting to see their first glimpse of daylight in 220 million years! We immediately christened the locality "Cycad Wash" and worked the quarry for the remainder of the afternoon. By the time we had finished, we had a backpack full of carefully-wrapped specimens, almost all of which appeared to be leaves of the same cycad. When the sun was getting low, we shouldered our hammers and packs, and set out to hike back to civilization, knowing that some of the fossils we were carrying would be good enough to be curated into the Petrified Forest museum collection. (Private collecting is not allowed in national parks.)

Back at the condo that evening, everyone exchanged stories and showed each other what we had found. The other crew had some nice teeth and scutes of phytosaurs and reports of awe-inspiring deposits of petrified wood in all the colors of the rainbow. We all felt privileged to see the backcountry areas of the park where vandalism has not yet removed all but the largest logs. We showed our finds to Vince and he selected the best specimens to keep. I was allowed to retain a couple of lesser-quality cycad leaf fossils to take back to the Denver Museum of Natural History for their paleobotany collections.

Petrified *Araucarioxylon* logs for which the park was named.

Vince also provided me with some literature from the park's library so that I could read up on the fossils I'd found.

The leaves were those of *Zamites* ("rock *Zamia*") *powelli*, a common member of the Triassic flora of the United States. It is thought to have been a deciduous plant of moist habitats. Though in macroscopic form it is impossible to tell the fossils we found from true cycad leaves, the microscopic structure of the epidermis places *Zamites* in the extinct group called bennettitaleans or cycadeoids. Both groups are gymnosperms, or "naked seed" plants, meaning that they bear their seeds in cones rather than in flowers, and are more closely related to modern ginkgos and conifers than to any other living groups. In fact, flowering plants were yet to evolve, and would not appear in the landscape until the Cretaceous. The cycadeoids resembled true cycads in growth habit and gross morphology, except that the structure of their cones was different (bisexual in some species, as opposed to male and female cones always being on separate plants in true cycads), and they bore them in the leaf axils, rather than at the apex of the trunk as true cycads do; this left conspicuous scars or pits on cycadeoid trunks. The stems from the former Fossil Cycad National Monument are actually those of cycadeoids, and you will frequently see them all called "fossil cycads" in the literature. Cycadeoids appear in the fossil record in the Triassic, whereas true cycads go back another 40 million years or so, being known from the Lower Permian. They are thought to have evolved from seed fern ancestors, probably some

Gary found this metoposaur skull near Tuba City, but we found another one on a reconnaissance foray in the park. Metoposaurs were large Triassic amphibians.

Aetosaur scute. Rock hammer for scale.

time in the Carboniferous. The closeness of relationship between the cycadeoids and the true cycads is still a matter of debate.

The remainder of our time was spent doing fieldwork in other areas, and on projects not pertaining to fossil plants. We discovered a beautiful metoposaur skull, and even found an arrowhead at a previously-unknown Paleoindian campsite—although we excavated neither, instead documenting them with photographs and locality info, and leaving them for another crew to collect. But for me, the time I spent looking for cycads was the highlight of the trip. Vince has now left the park, and we haven't been back, although WIPS has cooperative agreements with the Forest Service and the Bureau of Land Management and runs some very interesting trips to other areas each year. Perhaps some time in the future we will return to Cycad Wash, and unearth that cone that is still lurking somewhere just below the surface.

Arrowhead. Pencil for scale.

The Shirley Basin

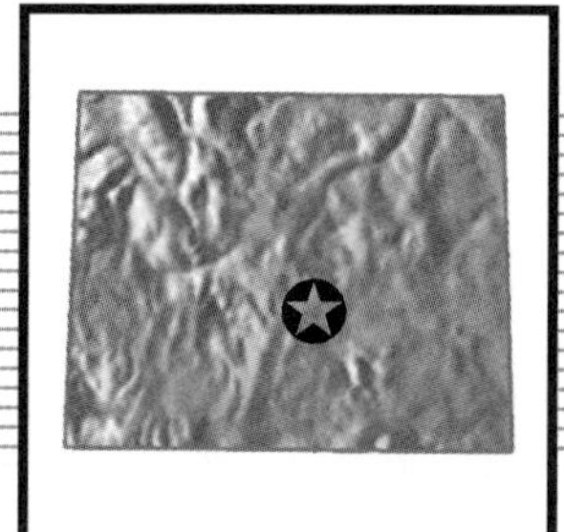

In August of 1994, my parents and I decided to take the kids up to Montana to see my sister and visit with their cousins. Earlier in the summer, my mom and I had signed up for a WIPS field trip to collect petrified wood in the Shirley Basin, but for reasons not remembered now, the trip had been cancelled. However, we had the

Wyoming shortgrass prairie with Laramie Range in background.

handout with directions to the locality, and it wasn't too far off our route...so who could pass it up?

The Shirley Basin area of Wyoming is northwest of Laramie and south of Casper, out in the middle of nowhere. The nearest berg is Medicine Bow, a one-horse town about 40 miles south of the collecting area. The basin is a typical Wyoming sagebrush prairie populated by pronghorns and a few rattlesnakes, some range cattle, and not much of anything else. To the east rise the mountains which separate it from Casper.

Pronghorns are extremely shy. My mom used a 10x zoom lens to capture these three.

The BLM land on which collecting is allowed exposes the Lower Eocene Wind River Formation. Most of the wood is probably of angiosperm (flowering) trees that grew in the subtropical or warm-temperate forests that covered Wyoming during that time. The landscape during the Eocene may have resembled parts of the southeastern United States today, with a thick, broadleaf forest. Early primates inhabited this forest, and some of their fossils have been found in other contemporary formations elsewhere in Wyoming.

The drive from Boulder to Billings can be done in a day, but not easily when you are travelling with small children, and not at all if you want to stop and have some time for fossil collecting. So we decided to make it a two-day trip to allow ample time for finding petrified wood, even though we had heard that there was so much wood to be found that you could fill the trunk of your car in an hour!

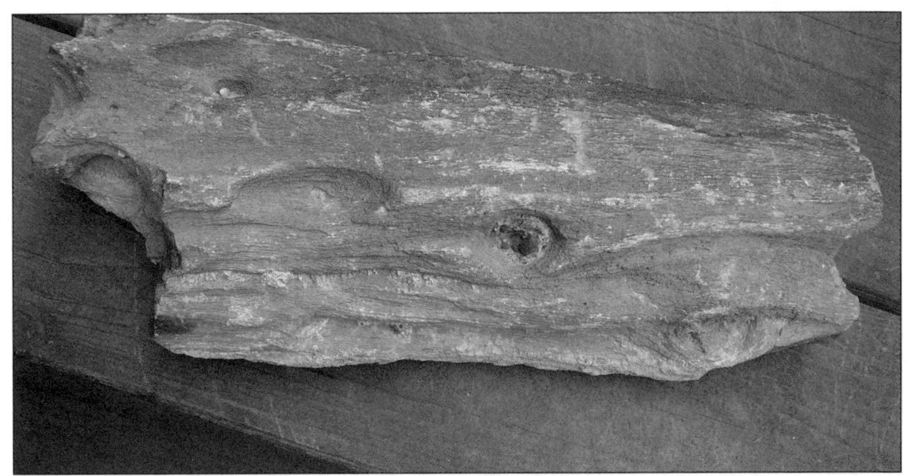

Mom collected this nice piece showing two knots.

After we found the site—no mean feat even with good directions—we were not disappointed. Pat, the original field trip leader,
had been right about how prolific the exposures were. There is as
much fossil wood lying all over the ground—not needing any excavation, just there for the picking—as there is in the backcountry of
Petrified Forest National Park, where the tourists haven't scavenged
all but the largest logs. The wood is a brownish-grey color; it isn't as
spectacularly colored as that in the Petrified Forest, but it preserves
a wonderful amount of detail, right down to the grain of the wood
and the presence of knots. There is one hazard, though (at least if
you are collecting at dusk, as we were): for every piece of petrified
wood you see, there are two cow pies which are about the same color.
So watch your step, and be careful what you pick up!

We didn't have a car with a trunk, but I think we could have
filled one in an hour if we had wanted to. You are allowed to collect
25 lbs. plus one large piece of petrified wood per person per day from
any one site on public land (up to a maximum of 250 lbs. per person
in a given year); and since there were five of us, we were entitled to
125 lbs. plus five large pieces that day. Of course, as dense as fossils

The side-by-side comparison of Shirley Basin wood and weathered, modern wood that I show the kids. Can you tell which is which? *

are, it doesn't take too many good-sized pieces to fulfill that limit; nevertheless, it's a lot of petrified wood. Some of the pieces are log-sized and too heavy to lift, unless you have several strong guys and a big pickup. I don't have room for those in my garden anyway.

Some of this petrified wood looks so much like modern wood, because of the fine detail, that you are hard pressed to tell the difference without hefting it. In fact, when I give a fossil talk to one of the local schools, part of my kit includes side-by-side comparisons of modern and fossil organisms, and one box contains an old piece of a weathered telephone pole that used to edge my garden planter—and a piece of Shirley Basin petrified wood. The kids guess right as to which is which about half the time.

* The longer piece on the left is the piece of an old telephone pole, and the piece on the right is the petrified wood.

12

Argentina:

the Dig that Almost Was

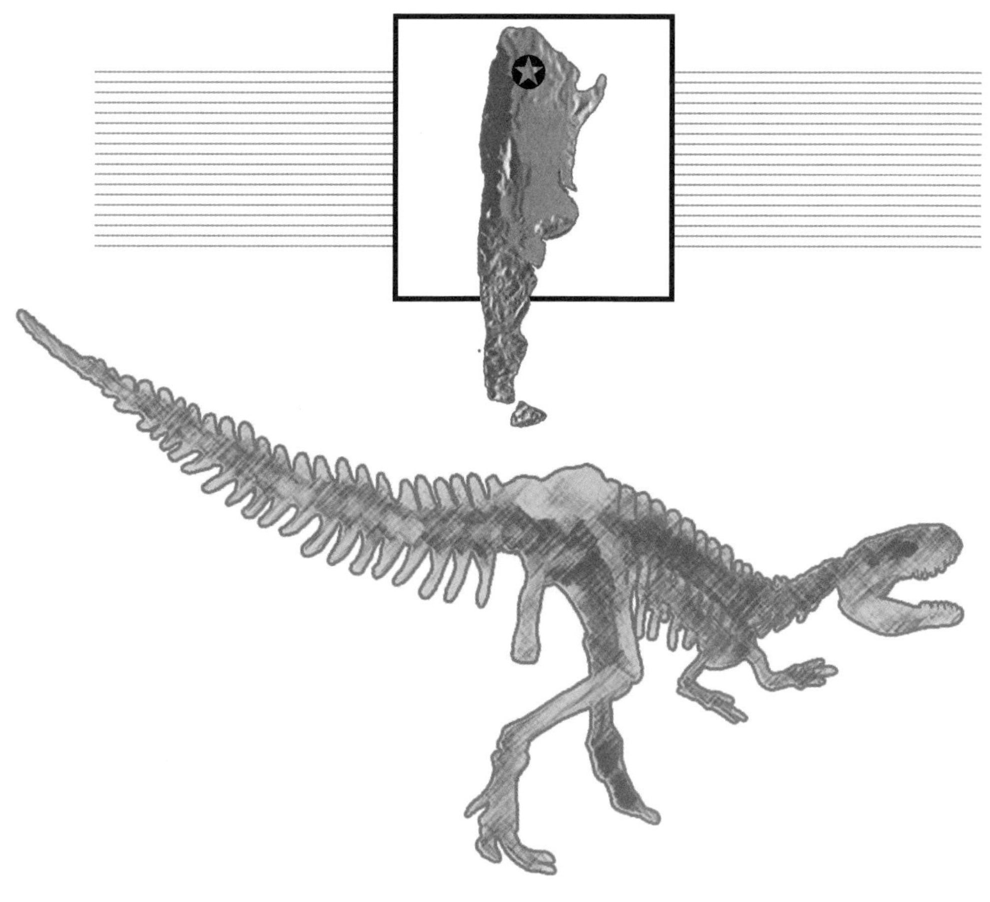

I n 1994, my friend Karen and I decided to go to South America for a month. She had never seen a solar eclipse, and very much wanted to catch the one that would be total over the Bolivian Altiplano on November 3rd. I had gone to Quebec with three girlfriends when I was 16 to view the total eclipse that occurred on July 10, 1972, and was game both for another one and the chance to traipse around South America unfettered. Mattie was just over a year old, and Allison was four, and I was getting kind of stir-crazy being cooped up with diapers 24 hours a day. Fortunately, since neither of them were in school yet, the time of year was not a problem, and I arranged babysitting with friends and grandparents so that I could be gone that long without Chris having to take time off work.

An acquaintance of mine put us in touch with a friend of his wife's, Rodolfo Aredes, who lived in Salta, Argentina, and was

Atacama Desert, northern Chile, the driest place on the planet. With less than 2" of rain a year, nothing—but nothing—grows here.

in the process of excavating some dinosaur bones there. In our halting Spanish, we wrote to him and arranged to include a couple days' digging in our itinerary...but as things turned out, it was not to be.

The trip started out flawlessly. Our plan was to meet—Karen flying in from California, me from Denver—in Santiago, Chile, spend a few days there, then work our way north through the Atacama Desert into Peru, visit Cuzco and Machu Picchu, and then head down to La Paz, Bolivia. We would rent a car and drive out into the hinterlands to catch the umbra of the eclipse, then travel to Argentina for the dinosaur dig, spend a few days with acquaintances of Karen's in Buenos Aires, and then work our way down to Punta Arenas at the tip of the Southern Cone, and back up to Santiago before returning home. We bought Lonely Planet guidebooks to prepare us for the hazards of unguided travel in third-world countries, packed our backpacks, and were off.

Alpacas at Sacsayhuaman ruins, Cuzco, Peru.

The cities in Chile and Peru were quite modern, moreso than we had expected, and up until Halloween, everything was going according to plan. The first hint of our troubles ahead came when we tried to check in at the airport in Cuzco for our flight to La Paz, only to discover that Karen didn't have her passport. No way were they going to let her out of Peru without a passport! We looked everywhere; our taxi driver went back and checked our hotel room; after an hour of having nightmares that we were going to be forced to reroute to

Lima and go to the U.S. embassy there to get a new one, the driver finally came back with the lost passport—it had fallen out on the floor in the back seat of the van. They held the plane for us and we had to dash out across the runway as fast as we could, enter the plane by the rear door, and collapse into the last two empty seats with the engines roaring. This little Halloween episode only presaged what lay before us in the days ahead.

We flew over Lake Titicaca and into the La Paz airport, which is at the edge of the 14,000-foot, treeless Altiplano. La Paz itself sits in a bowl where the Altiplano breaks away towards the lowlands of eastern Bolivia. It's a very picturesque city, with eucalyptus trees lining cobblestone streets, and we were staying in a 17th-century guesthouse with white stuccoed walls and a red tile roof, seemingly right out of old Spain. We thought our troubles of the day were behind us, until I went out in stocking feet to take a picture of the city from our

Machu Picchu.

balcony, and slipped on the polished terra cotta steps. My feet went totally out from under me, and I came crashing down hard right on my backbone on the unforgiving steps before I had the chance to put out a hand to catch myself. The pain was so intense that I feared that I had fractured a vertebra or pelvic bone. I could limp along slowly using a walking stick, but couldn't get out of bed without Karen's help, or get in or out of a chair without clutching my stick with all my strength. Had I been back in the States, I'd have gone right in for an x-ray, but I was nervous about go-ing to a South American hospital and trying to explain the problem in the minimal amount of Spanish I knew. So I put up with the pain and hoped for the best, deciding that I was in no immediate danger since I could walk, and that I'd give it a few days and see how things went before deciding if I needed medical attention.

Yet we were undeterred in our plans to see the eclipse. We rented a truck on the morning of Novem-ber 2nd, and Karen volunteered to drive. Not only did I have my back problem, but the "rules of the road" in Bolivia are something else—few intersections have stoplights or stop signs, and drivers merely honk their horns before speeding through to let others know to get out of the way. I breathed a sigh of relief when we got out of La Paz and onto the sparsely populated Altiplano, where the traffic was much thinner and the road crossings fewer.

The second native family we gave a ride to.

The band of totality was going to pass south of Oruro, which is maybe 120 km from La Paz. We would have to drive 50 km out onto the salars (salt flats) and camp there, because the eclipse would occur around 8 o'clock in the morning. Well, what you get when you rent a car in Bolivia is not the same thing as when you rent a car in the States. Our vehicle, a white pickup, had over 100,000 miles on it and had definitely seen better days (a lot of the vehicles you see on the road down there would be found at junkyards in this country). But it ran—when we got it, that is. Driving a pickup garnered us hails for a ride from a couple groups of locals, which we were glad to provide. They would flag us down from the roadside and the whole family would pile in the back, then knock on the driver's window when they wanted to be left off. The Altiplano landscape looks a great deal like portions of our Great Basin, with rolling, grassy flats ringed by high mountains. Towns along the road are just clusters of adobe huts. Karen has a healthy appetite and ate some concoctions at a small cantina in a nameless town that I decided I wasn't hungry enough to try. A persistent little boy there was peddling some concretions containing nice trilobites that presumably had been found nearby. Altogether it was fascinating, and I always figure, if you want things to be just like home, you ought to stay home.

So there we were, about 20 km from Oruro, when the engine started to overheat. We pulled off on the shoulder and opened the hood, only to discover that the fan belt had broken. What to do? We didn't have a spare fan

Trilobite in concretion.

belt; it would take forever to walk to Oruro; we couldn't drive that far on an overheated engine without boiling the radiator dry, and ruining the engine altogether. We had pretty much resigned ourselves to the fact that we were going to have to sit by the roadside until the engine cooled, drive slowly as far as we could till it overheated again, and just repeat the process until we could limp into Oruro that way. So much for any chance of getting out onto the salars to camp that night, but maybe we could get a motel in town and get up early...but after this, did either one of us want to trust this vehicle cross-country, especially with me barely able to walk?

Just when you need a penny from heaven, one arrives in the form of two truckers driving a semi into Oruro. The sight of two ladies, blonde and obviously foreigners, stopped by the roadside with their hood open, brought out the kindness in these two burly guys, and they stopped and asked if we needed some help. Through phrases and gestures we explained to them what our problem was, and they proceeded to fabricate us a temporary fan belt out of a piece of rope I had tied around my backpack and some wire. Then they drove behind us at 20 kph all the way into town to make sure we got there all right. And that second local family we had riding in the bed of our pickup? The father knew right where an auto repair shop was. The reason the fan belt had bro-ken was because the alternator had seized up, but we got both of them replaced pronto for a grand total of a little over 5 bucks! I kept our "Bolivian fan belt" as a souvenir of our little misadventure, knowing no one back home would ever believe the worth of that dirty little piece of cotton rope.

Frayed and grease-stained, this little
piece of rope saved our skins.

By the time we had gotten our truck fixed, it was starting to get dark. I was willing to get a room in Oruro and get up at 4 a.m. to drive out to the umbra of the eclipse, not because I wasn't nervous about it, but because I knew how important seeing the eclipse was to Karen. But when she declared that she'd had it with this vehicle and would rather drive back to La Paz that night, get a really nice hotel, and see a 97% eclipse from civilization rather than risk our safety to that truck to see totality, I didn't need much convincing. We certainly would be in a bad way if the truck broke down again out on the salars, especially with my back problem, and together we decided it just wasn't worth the risk. I had seen a total eclipse before, so I could deal with missing this one if she could. We called ahead and reserved a suite at the best place in La Paz, made it back without any more troubles, and settled in to a nice dinner (of food I could recognize!) and some maté de coca to calm our frazzled nerves.

Well, wouldn't you think that was enough adventure for one trip? Think again! To get to Salta, we had to fly from La Paz to Yacuiba, a small town on the Bolivian side of the Bolivia-Argentina border. We expected Rodolfo to meet us at the airport, but he was nowhere to be found. And to top it all off, Karen's luggage wasn't there when they unloaded the plane—and her prescriptions were in the lost bag. The airline assured us that there was another flight from La Paz due at 5 p.m. and that they had found her bag and it would be on that flight, but we would believe it when we saw it. There aren't any phones in Yacuiba, so we had to walk across the border to Pocitos to call Rodolfo, and found that he was home watching TV. Something had gotten mixed up in the translation, and he thought we were going to take the bus from Pocitos to Salta. Well, we could do that, but we had to wait for Karen's luggage—which was going to be all day. Remember how I told you that the cities were more modern than we had expected? Well, Yacuiba is your third-world nightmare town, seemingly right out of *Romancing the Stone*: hotter than the hinges of Hades (it's in the lowlands), humid as all get out, jungle all around, unpaved streets with potholes the size of Volkswagens, and goats wandering about town past stands of rotting fruit. (To this day, when Karen and I joke about someone who deserves a booby prize for something, we say we'll give them an all-expense-paid vacation to Yacuiba.) Karen is pretty easygoing most of the time, but she gets very cranky when she gets hot, and the lost luggage didn't help. She also gets hungry easily, and we had spent practically all of our bolivianos, figuring we would need pesos in Argentina, so we couldn't even buy lunch—and the cambios (money-changing houses) take a siesta from 12 to 2, so our American money didn't help. My back was giving me as much trouble as ever, so I couldn't wear my backpack, and she had to drag it for me, adding to her crankiness. Again, I was willing to wait for the luggage and then board the late bus for Salta, but she decided she'd had it with the hinterlands and wanted to get back to civilization and air conditioning. It would mean giving up on the dinosaur dig, but I wasn't sure I'd be able to scrunch down in a pit or shovel

Hall of prehistoric life, Museo Argentino de Ciencias Naturales, Buenos Aires.

loads of overburden in my condition anyway, so when she declared that she wanted to take the late flight to Santa Cruz and then proceed on to Buenos Aires, I didn't give her an argument. It was disappointing to have come all that way and not get to dig dinosaurs with Rodolfo, but sometimes you just have to throw in the towel.

Things were looking a lot brighter after a good night's sleep in an air-conditioned villa in Santa Cruz. It had been a week with no improvement since I had hurt my back, and I had pretty much decided to get an x-ray once we got to Buenos Aires, but then suddenly overnight my back pain improved tremendously, and I could begin walking without a stick. It must have just been a bone bruise after all. In Buenos Aires we stayed at the home of one of Karen's fellow alumni from a small college in California, who just happened to be a travel agent, so he put us onto some things to do in the city and the rest of Argentina. I always insist on visiting the natural history museum in any big city I go to, so even though we didn't get to dig for any dinosaurs, we did get to see unparalleled mounted specimens of Argentina's prehistoric inhabitants at the Museo Argentino de Ciencias Naturales (Museum of Natural Science). It is a large stone building with several floors, and most of the ground floor is devoted to exhibits on Argentina's past, including an impressive array of mounted skeletons. It has a dusty, turn-of-the-last century look to it, and reminded me of the Musée National d'Histoire Naturelle in Paris, except that it didn't have all the jars of pickled creatures that the French museum does.

Magellanic penguins,
Valdes Peninsula.

We also visited another, smaller but very nice, museum in Trelew—the Museo Paleontologico Egidio Feruglio. We had gone to Trelew to visit the nature sanctuary on the

Elephant seals, Valdes Peninsula.

Mount of fossil mammal, Museo
Paleontologico Egidio Feruglio, Trelew.

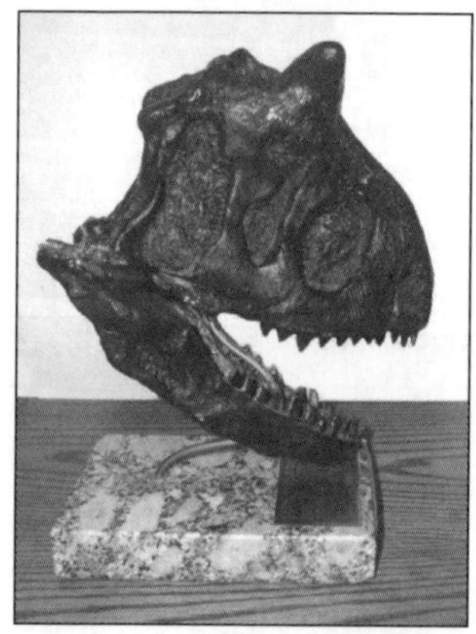

Carnotaurus skull cast.

Valdes Peninsula, and see the penguins, whales, and elephant seals. I did notice some fossil clams embedded in the sandy cliffs, but there was no one I could ask about the age or origin of the sediments. They were poorly-consolidated sandstones that looked to me to be probably Quaternary and likely no older than Tertiary. I don't remember seeing any information on the geology of that area at the paleontological museum. On the other hand, this small museum had fabulous exhibits on Patagonian dinosaurs, including nests of eggs and some great murals on prehistoric Argentina. Some of the fossil material came from the famous deposits at Ischigualasto, which we were unable to visit, and other material was from elsewhere in Patagonia. I couldn't resist purchasing a beautiful cast of a perfect *Carnotaurus* skull for only 50 pesos ($50). This midsize theropod is only known from Argentina and gets its name from the two small horns adorning its skull (its name means "carnivorous bull"). It is classified in the Abelisauridae and is known from Cretaceous deposits. Some scientists have theorized that the horns may have been used in head-butting contests like those seen today in bighorn sheep. The skull cast is mounted on a polished granite base, and I display it on a bookshelf in my library. I managed to get it home with only a single broken tooth!

That was the end of both the paleontology and the misadventures of our trip. We climbed a glacier at Lago Argentina, where we hiked through a forest of *Nothofagus*, the Southern beech, a remnant of a southern Gondwanaland flora. We spent a couple of days in Punta Arenas (the southernmost city on the mainland, right across the Strait of Magellan from Tierra del Fuego), and stopped at a couple of other places on the Chilean coast before returning to Santiago and home. What with all our mishaps, you might think I couldn't wait to get back to the States, and I have to admit, after a month, I missed my kids—diapers and all. But by now, I'm ready to do it again if I ever get the chance, and this time I'll know better than to walk on terra cotta steps in stocking feet!

Perito Moreno glacier, Lago Argentina. I missed the dinosaur dig, but my back was enough better by this time for me to don crampons and climb this ice massif.

And at least I have one total eclipse
to remember—

Solar Eclipse in Quebec, July 10, 1972.

On a sand bar, by a river,
In a valley lined with pine trees,
In the west the sun shone brightly,
Round and full, a dragon's bait.

From the north the beast approaches,
Hid in mist, the sun as bright;
Takes a nibble from his belly,
Bites again with waxing greed.

Subdued by mist, the sun is prey
To be devoured bite by bite.
With bloated gut and coal-black skin
The monster burps and closes in
To eat the sun's last light.

The sky is night, the valley hushed,
The black and evil thing
Is broken by a rim of light—
A dancing fairy-ring:
The placid moon slides off and south.
It opens wide its gluttoned mouth
And lets the sun slip through.

13

Blue Mesa Sauropod Rescue

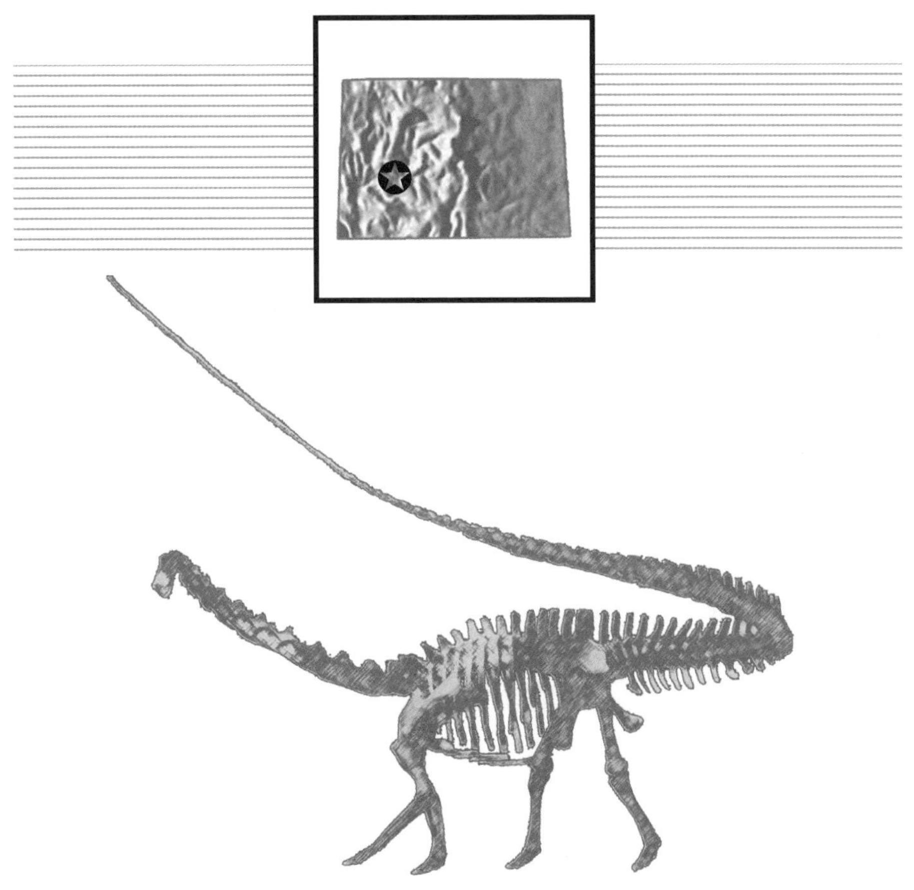

Exposures of the Morrison Formation are widespread in the Western Interior of the United States. They consist of a distinctive series of multicolored sandstones, mudstones, and shales which were deposited on low-lying floodplains and in fluvial environments during the Late Jurassic Period. These sediments are famous for the dinosaurs they contain, notably including those at Dinosaur National Monument, which straddles the Colorado-Utah border.

Blue Mesa Reservoir as seen from the dig site.

My involvement with the Morrison this time was as part of a team from WIPS which was assisting the National Park Service in retrieving a sauropod which was about to drown. (Well, okay, if drowning was the cause of death, it happened 140 million years ago, but now the fossilized bones were about to be submerged under the waters of Blue Mesa Reservoir.) The Park Service wanted us to get out as much of the beast as we could before the reservoir swelled to its highest level in many years, replete with meltwater from the record snows which El Niño brought to us in Colorado in the spring of 1995.

Don and I are jacketing bones.

As digs go, we had it pretty cushy on this one. The Park Service let us stay in a dormitory-style condominium (cooking facilities, hot showers at night!) and taxied us out to the site by boat each morning. The bones were embedded in the hillside near the shore of the rez, and at the time I was there, were a good thirty feet above the water level. We disembarked from the boat, climbed the side of a small gully, and were there. No marching miles across searing desert with heavy backpacks: these bones lie at about 7000 feet elevation in the Rocky Mountains, near Gunnison, Colorado. The scenery is spectacular and the weather was perfect for digging—coolish and overcast, but not much rain. When the occasional shower came we all retreated under the "Fred shed," a tarpaulin-covered shelter named after one of the dig's leaders. We even had a porta-john available; now is that luxury, or what?

Until the bones are all extracted and more fully studied, it is only possible to say that they are sauropod remains. That's one of those long-necked, long-tailed dinosaurs, like *Apatosaurus* ("Brontosaurus") or *Diplodocus*. Which one, no one is sure. And we didn't even know how much of the skeleton was there. We worked on vertebrae, ribs, and something dubbed the "UFT" (unidentified flat thing). The bones are in a matrix of mixed sandstones and mudstones typical of the Brushy Basin Member of the Morrison. This means that in some

Series of articulated sauropod vertebrae.

places, the surrounding rock is quite soft and easy to remove, and in other places, it's as hard as...well, rock. Hammers and chisels were standard equipment, along with plenty of Vinac (preservative), and superglue to repair damage (some done by Mother Nature, some done by us). Many of the bones needed plaster jacketing to hold them together. Sometimes, we could remove the bone from the rock, but the pieces didn't fit together well enough to be glued and had to be held in juxtaposition by the burlap and plaster. Other times, the bones were too fragile to be removed from the matrix in the field, so we had to jacket large blocks for removal. Such was the case with a series of several articulated vertebrae which extended into the hill-side.

Jacketing bones in the field is a really interesting process. It hasn't changed much since it was first used by the old-time prospectors in the latter half of the 19th century. It's more or less the same process as when the doctor puts a cast on your broken leg: strips of cloth (in this case, burlap) are dipped in plaster of Paris, and then used to cover the bone papier-maché style. Sometimes fragile bones are jacketed in place without being fully excavated—other times, we removed them from the rock, and then jacketed them to hold them together for transport. The jackets are later removed back at the museum or lab, and the bones stabilized permanently with glue.

I worked most of the time removing matrix from around the vertebral sequence with a hammer and chisel. My mom, who accompanied me on this trip, worked on a rib. You should see my mom excavate dinosaur bones—she's the most meticulous worker I've ever seen, and it takes her forever, but a fossil bone couldn't ask for a better caretaker. There won't be a smidgen of damage done to that bone, and it'll be clean as a whistle when she's through. I bet the preparators back at the lab wish that everyone would be so careful!

My mom's rib, extracted and ready for jacketing.

We would have liked to be out there digging day after day, but other commitments dictated that my mom and I be just two of several members of a rotating crew. The short time we spent at Blue Mesa was a lot of fun, though. (My only complaint was of the abundance of ticks, and a little superglue froze those in their tracks: fossils in the making). Two weeks later when I talked to Doug, the other team leader, he told me that the water level was rising so fast that he expected the Fred shed to be underwater that afternoon. Another ten feet—not an impossible rise in a day when the snows melt—and the bones would be flooded. Some of the blocks we removed were estimated to weigh over two tons, and it was touch and go whether they could be safely loaded on the Park Service boat. Doug was trying to get through to Fort Carson to see if he could borrow a helicopter so he didn't have to winch the blocks up the hill to avoid the rising waters. (Fortunately, as it turned out, the boat did manage to hold even the largest block.) As for the rest of the sauropod, well, he's blowing bubbles by now, waiting for a crew to return again when the water is lower sometime.

14

The Real McCoy

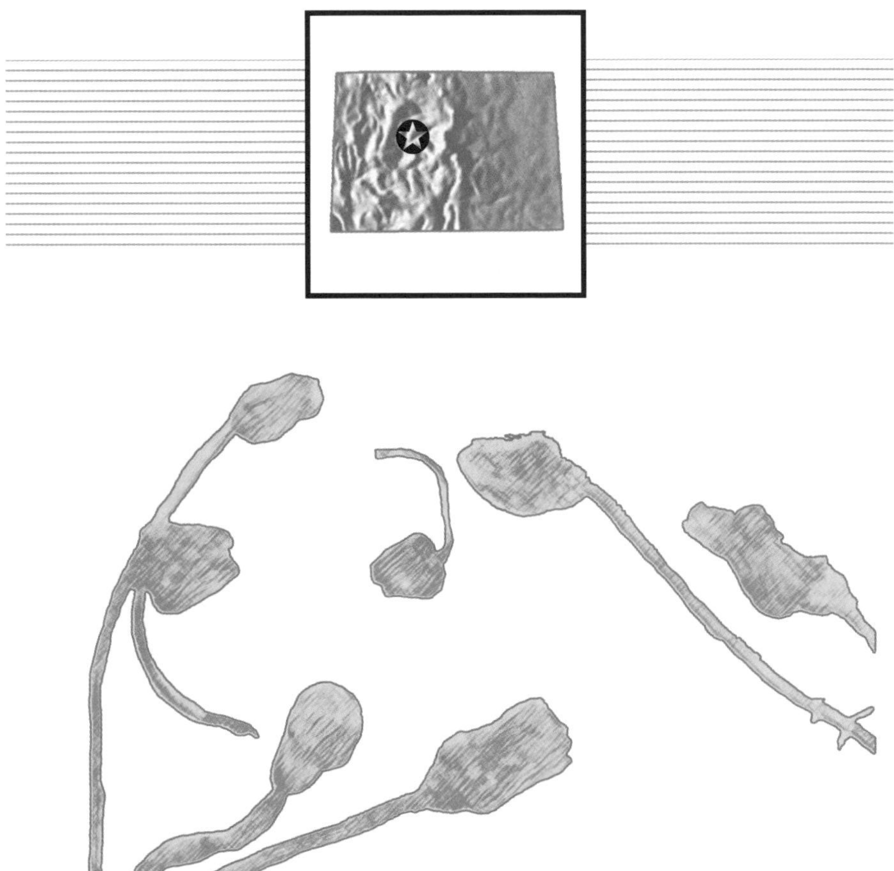

W hen the Laramide Orogeny pushed up the Rocky Mountains during the Late Cretaceous, large intrusions of igneous and metamorphic rocks formed their core. In much of Colorado's mountainous country, these rocks are exposed at the surface, and your chances of finding a fossil are nil. That's the case along most of the Front Range, where all the sedimentary layers which originally topped these mountains have long since worn away.

But on the Western Slope—a term which locally means "everything on the west side of the Continental Divide"—many of these sedimentary rocks remain. Vast exposures of what are called the "Penn-Permian redbeds" (named, obviously, for their age and

The crinoid gully at McCoy.

color) can be found in the central mountains of the state. One area dear to collectors is the Middle Pennsylvanian Minturn Formation in the vicinity of the one-horse town of McCoy.

I've been to McCoy several times, first in 1992 on a WIPS trip led by Wayne Itano, and again in the summer of 1995, collecting for the University of Colorado under the supervision of Karen Houck. Karen's Ph.D. research was on the fossils from this area, and we were there to assist her in collecting specimens and doing stratigraphy for her thesis.

The stratigraphy of the area is quite complex, and is further complicated by intense faulting. During the timeframe when the Minturn was deposited (about 306-303 million years ago), the area lay along the eastern flank of the Central Colorado Trough, a northwest-southeast trending basin formed by the collision of South America with North America. Block-faulted mountain ranges flanked this trough, which was located only about 15° from the equator. A series of marine transgressions and regressions deposited various nearshore marine and alluvial fantidal flat sediments in the area. This means that depending on where you are collecting in the formation, you might find shark teeth or spines, brachiopods, crinoids, echinoid spines, clams, or even some terrestrial plants. We visited several locations and collected representatives of most of these.

Various types of crinoid stems from McCoy.

Although McCoy is in the high country at perhaps 7500 feet, it can still get hot in July. Sunscreen is essential, because the thin air filters out few of the ultraviolet rays. I have to be careful about getting too hot because that is a sure trigger for one of my migraines. So after we spent the morning measuring a strat section out in the open, I sought out some shade when we made our first collecting stop. It was in a small gully that trended northeast, and if you climbed up it a ways, juniper trees provided small spots of cool shade.

Small fern leaf.

This particular area had more disarticulated crinoid columnals than any place I've ever seen. Crinoids were ubiquitous stalked echinoderms of the Paleozoic, which are nicknamed "sea lilies" because they looked more like plants—resembling a tulip without leaves—than the animals that they were. Although a few species inhabit the ocean floor today, crinoids never regained the zenith of their diversity after being decimated in the great extinction at the end of the Permian.

Whole, articulated crinoids are rare, and are never found in the Minturn. The segmented stalk was made up of a series of round plates, resembling small buttons with a hole at their center, strung together like one of those candy necklaces. When the crinoid died, the stem usually fell apart and the segmented columnals were scattered across the seafloor. Worldwide, these crinoid columnals are some of the most often-collected fossils there are. Kids love them, and since you are allowed to collect reasonable (non-commercial) quantities of invertebrate fossils on public lands, I brought back a whole slew of them to give to the kids at the schools when I give a talk. The vast majority of these fossils have no scientific importance, and the area has been extensively studied, but later, when she was at the University of Kentucky, Karen published a description of a new type of crinoid from McCoy which she dubbed *Sciadiocrinus wipsorum*, in honor of WIPS and the help the members have given her.

The following day we collected at two different locations. At one, a spot atop some rolling hills with lots of nice shade from juniper trees, the sediments were grey rather than red, and we found lots of brachiopods and a few nautiloids. The Minturn is a complex formation with 19 named members of varying lithology—conglomerates, sandstones, shales, and limestones. Even within these members there are a plethora of distinctly recognizable layers, and if I ever knew, I've forgotten just where in the formation this spot was. The same holds true for the last spot we visited. It was on the side of a steep hill under more junipers, where an older sandstone—I don't recall the exact age—was exposed. From here I brought home a nice specimen of some early land plants. Were you not to look quite closely (or have someone along who would point them out to you), you might easily overlook these fossils as merely iron stains in the rock. They resemble brown pipe cleaners more than anything else, but in reality are the fossils of simple, stalked plants that had yet to develop true leaves. The sideways projections coming off the central stem are

Early "pipe cleaner" plants.

enations or proto-leaves. My fossil collection isn't large compared to some people's (like Bryan Cooney, whose entire basement is filled with shelf upon shelf of small cabinets containing drawers of fossils, all neatly labeled—almost like the collections of a museum!), and it isn't that well organized; the majority of my specimens are still in ziplocs or Pringle's cans with little notes stuffed in as to locality and date, all shoved in the drawers of a large old dresser in my basement. But my best specimen of these early "pipe cleaner" plants has an honored spot on an end table in my library.

Brachiopods from McCoy.

Someone from WIPS usually runs a trip to McCoy almost every summer, and since it is on BLM land, you are free to go out there and collect any time you choose. This summer my mom and I took Allison and Mattie up there so that they could collect their own treasure bags full of those little crinoid columnals. We collected at that same small gully that Karen had taken us to in '95, and the fossils were as plentiful as ever. You go a couple of miles west of the highway on the gravel road to Burns, and stop right next to a very large ash tree. The gully is on the east side of the road opposite the big tree. We found half a dozen different kinds of crinoid stems and a handful of very nice brachiopods that I don't remember having collected the last time I was there. The kids were really impressed, because they had never collected these types of fossils before, and since the crinoids are very well-preserved, it doesn't take a highly trained eye to recognize one. Mattie was excited when she

Mattie's lizard.

Yuccas and globemallows at McCoy.

spotted a small lizard scurrying among the rocks, and even took a picture of him when he stopped to rest under a clump of sagebrush.

By this year, my mom's arthritis precluded her from climbing uneven terrain, so she didn't do any collecting of her own. But her special interest—photographing wildflowers—kept her busy while the rest of us girls scrambled around looking for fossils. It was early June and the mountain lupines were in full glory, along with purple penstemons and larkspurs; but near the collecting spot, the show was stolen by the flowering of yuccas and salmon-orange globemallows—so a good time was had by all.

15

Florissant Fossil Beds

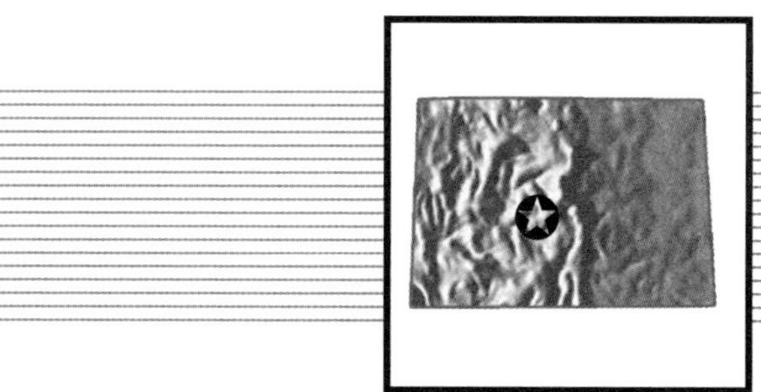

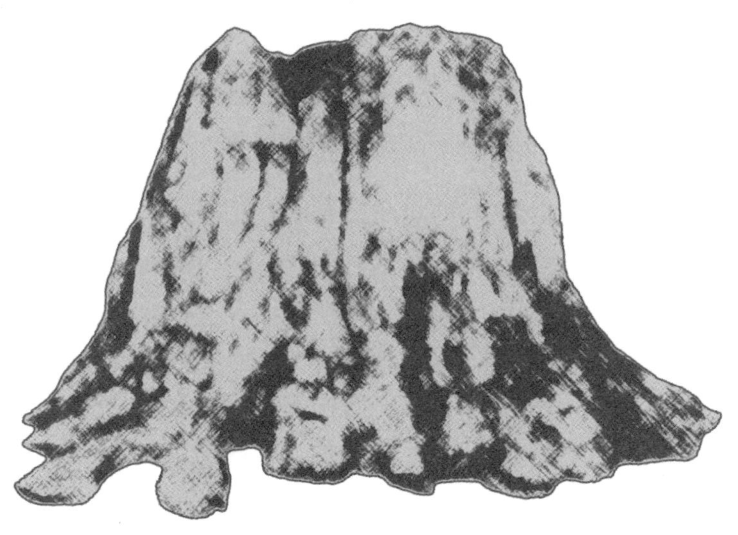

Florissant Fossil Beds National Monument is located in the mountains of central Colorado just west of Colorado Springs. It is in a high valley sandwiched between the Pike's Peak massif and the hills ringing South Park. Grassy meadows alternate with pine forests, and in July, the parks are ablaze with wildflowers: scarlet Indian paintbrush, blue-purple harebells, and every type of yellow composite you can imagine. (A French word for 'flowering' is the

The shale at Florissant is very soft, and I found that my artist's palette knife was the best tool for carefully prying the layers apart.

etymology of the name Florissant.) Add to this perfect temperatures and a crystalline blue sky, and you have the setting for a delightful paleontological expedition.

We were working in the northwestern part of the park, assisting the National Park Service in attempting to relocate some old quarries and excavate them for specimens to add to the museum's collection. The work was directed by Herb Meyer, the park paleontologist. Herb had contacted WIPS to solicit volunteers, and that's how I got in on the dig.

The fossil-bearing layers are part of the Eocene Florissant Formation, a volcanigenic unit about 34 million years old which sits atop the Wall Mountain Tuff, a welded tuff capping the billion-year-old Pike's Peak Granite. The Florissant Formation consists of several volcanically-derived units, some of which are lacustrine sediments and others of which are mud and debris flows. At the time of its deposition, the Rocky Mountains had already been uplifted and a stream

Fossilized *Sequoia* stump.

My mom enjoys fossil digs that entail careful work and not a lot of walking.

Fagopsis, an extinct relative of the beech.

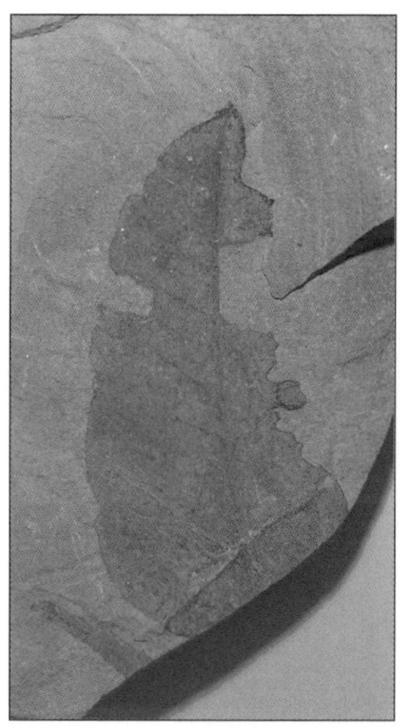

Angiosperm leaf, possibly something related to a western sumac.

valley cut in the present location of the park. A *Sequoia*-angiosperm forest growing in the area was covered by a mudflow that dammed the valley to the south when a nearby volcano erupted. Some of the large *Sequoia* trees were preserved *in situ* as standing trunks. A lake formed behind the dam, and leaves and insects fell into the lake and were fossilized in the fine sediments. This sequence of events was repeated, resulting in the deposition of some wonderful fossil-bearing layers. Although a few vertebrates and some snails have been found, it is the plants and insects which have made Florissant famous and for which the national monument was established.

We had to walk about half a mile along an old jeep trail to reach the quarrying area. The sediments there are unbelievably fine, and rock hammers alone are not appropriate tools. We found the most useful things to be small chisels (for prying), bricklayer's scrapers, and palette knives. The shale disintegrated into cardboard-thin layers which had to be carefully removed from the hillside and inspected for fossils. Many times a piece which appeared uniformly grey when fresh would yield compression-impression fossils of beautiful leaves when it dried a little. Then we'd use a nippers mounted to a 1x6 to trim away the extra rock from around the fossil. If the piece was very thin and fragile, Herb showed

Part and counterpart of a compound leaf of uncertain affinity.

us how to use a little Elmer's to glue the playing-card-like piece to a heftier piece of shale for a backing.

Most of what we uncovered were small leaves of angiosperms (flowering plants and trees), such as *Fagopsis*, an extinct relative of the beech. There were also pine needles, white cedars, rose leaves, elms, one pine cone, and a bug or two. Although we didn't of course find them all, over a hundred types of leaves are known from the formation, including oaks, willows, sycamores, ferns, horsetails, sassafras, mulberry, walnut, birch, hawthorn, apple, maple, sumac, and many others, some representing completely extinct groups. Most of the specimens

Allison proudly shows off the leaf she found at Claire Ranch.

were very well preserved by the fine sediments and showed the details of leaf venation upon examination with a hand lens. Herb told us that the pattern of leaf veins was very important in diagnosing the taxon to which a leaf belonged. Some two hundred types of bugs, from flies to butterflies to spiders, also come from the same sediments, but they seemed to be hiding while we were there. We carefully wrapped all of the specimens Herb wanted to keep in toilet paper and placed them in pop flats before loading them in the truck at the end of the day.

Allison's hickory leaf.

My mom and I were only there for one weekend, although Herb had volunteers helping during weekends throughout July. Being a national monument, all specimens are government property, so we couldn't keep any of the fossils that Herb didn't want. But right outside the park is a privately operated quarry, Claire Ranch, where you can dig through shales of the same formation for a small fee and keep what you find. We went back there with my five-year-old daughter, Allison, late in September, and she found the best leaf of all: both halves of a four-inch specimen of the hickory *Carya libbeyi*. We had great fun, both digging for the park service and at the private quarry. And you can't beat the scenery!

16

Big Cedar Ridge

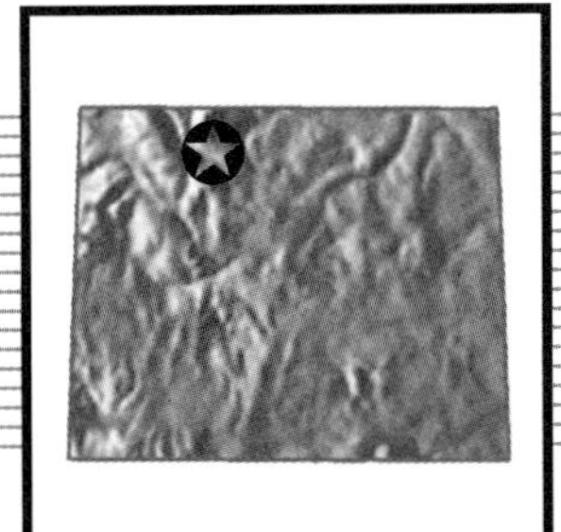

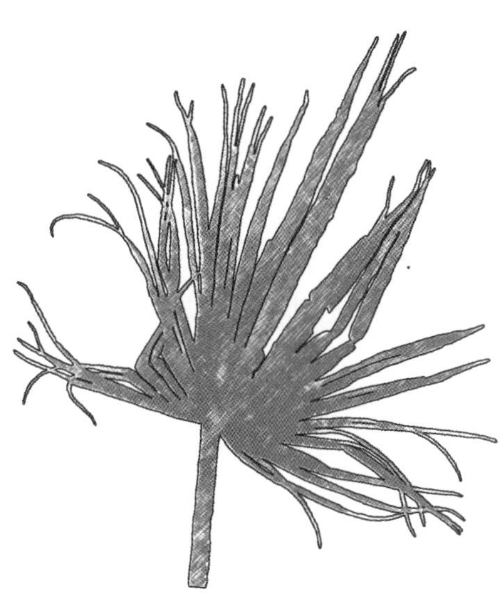

If you like milkshakes (and I do, with double malt), this was the trip to go on. The dig site was located just minutes from Dirty Sally's in Tensleep, and our leader, Kirk Johnson, made sure we didn't work too long without recharging our batteries. Happy workers find more fossils!

Big Cedar Ridge is a unique site in the Meeteetse Formation of northern Wyoming that preserves a beautiful snapshot of the

Big Cedar Ridge.

The shade canopy made digging pleasant in the hot sun.

botanical communities of the Late Cretaceous (71 million years ago). The Meeteetse is sandwiched between the Lance Formation and the Lewis Shale in this part of the Bighorn Basin. It is recognizable at the surface by its grey color and popcorn-like texture. The bentonitic tuff, a remnant of ashfalls from volcanoes in Montana and Idaho, covered everything and fossilized the plants *in situ* and in three dimensions. For this reason, Big Cedar Ridge is called a "Plant Pompeii." Whole plants can be found instead of just the detached leaves common at other sites. Herbaceous vegetation is preserved here, while it is not frequently found at other paleobotanical locations.

The paleobotanical community here varies with the paleoterrain and is represented by one of three characteristic assemblages: ferns and palms growing on an organic silty substrate, small angiosperms (flowering plants) in disturbed areas, and a fern-cycad flora on peaty soil. We were split up into small groups, all working different sites along the east face of the ridge. I requested a cycad site and was assigned to quarry 13 (maybe that's why I never found any cycads, although it was a productive locality for other types of plants). We were collecting for the Denver Museum of Natural History, and after Kirk inspected our finds and took what he wanted, we were allowed to keep what was left.

Our spot was on the nose of a small knoll jutting east from the main line of the ridge, which runs north-south and offered a little bit of protection from the strong westerly breeze. In fact, most of the time, conditions were very nice for digging. It was sunny, probably in the mid-80s, and breezy. The first day it might have been a little hotter, but our shade canopy held up for most of the day before succumbing to a particularly strong gust. The wind kept it from really being too hot. The scenery was beautiful: multicolored badlands and not a soul around for miles except the other members of our crew. It was nice and quiet, so we got to drink in the blissful solitude for hours. For me, it was especially refreshing, because my dad kept the kids entertained at the campground in Worland while my mom and I looked for fossils.

We found one small pinnate frond that Kirk told us was a type of conifer, and some fragments of palm fronds, but most of the palms came from sites further along the ridge where other members of our crew were digging. Our best find was numerous examples of a type

Floating ferns.

of fern which had radially arranged leaflets, like an oxalis, but which resembled the pinnae of a maidenhair fern. Kirk told us it was a floating fern which inhabited the surface of pond areas like modern duckweed does. I guess that's why it's associated with a peaty substrate, which may have been what we were calling hash layers, bedding planes just covered with thousands of minute plant fragments, so many that they colored the layer a darker grey than the rest of the rock.

Studies of the Big Cedar Ridge flora have shown that angiosperms (especially dicots), while they were extremely diverse, constituted but a small fraction of the plants in the overall flora at the time. They were not yet dominant over the ferns and gymnosperms that held sway earlier in the Mesozoic. In later times, they would paint the landscape with the bold splashes of their flowers, but 71 million years ago at Big Cedar Ridge, things were still mostly just green. The only angiosperms present in any quantity were the monocotyledonous fan palms.

Fan palm fronds.

The weekend ended too soon, as fun times always do. We continued on up to Montana to visit my sister, while most of the rest of the crew headed back to Denver. The museum got a lot of good fossils, and I even came home with some nice pieces of ferns and palm fronds to add to my collection. Thanks, Kirk!

17

Bonanza Paleobotanical Bonanza

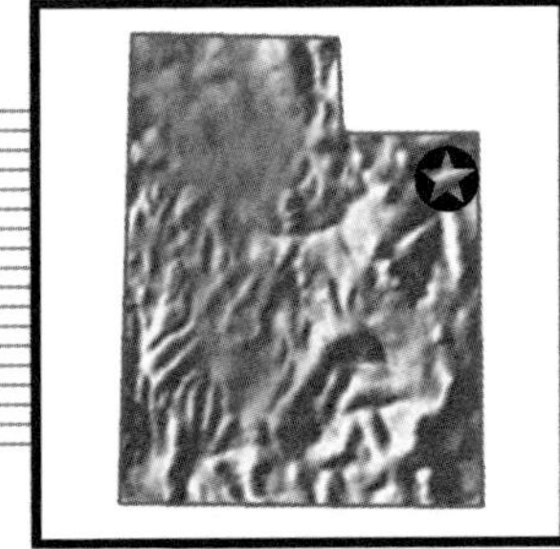

During the winter/spring of 1996, I had the opportunity to take a paleobotany class that Kirk Johnson was teaching at the Denver Museum of Natural History. When I was in grad school at the University of Colorado, there weren't any paleobotanists on staff, so I had never taken any paleobotany. I did take botany as an elective when I was in engineering school, and you know by now that I've grown a lot of houseplants and have a special interest in cycads, so I jumped at the chance to find out more about fossil plants than I had heretofore soaked in.

We camped here next to the White River.

As an adjunct to the class, in late April we had the option of going out to the miniscule berg of Bonanza, Utah and helping Kirk excavate a site he'd found on BLM land. Bonanza is in the northeastern part of Utah, just a few miles from the Colorado border and near to the booming metropolis of Dinosaur, Colorado (gateway to Dinosaur National Monument). April usually has fairly decent weather in this part of the country, and the elevation wasn't particularly high (~5500 feet, like all of the "lowlands" on the Western Slope, but not high in the mountains)—however, springtime in the Rockies is always unpredictable, and for this trip we drew 42°, windy weather.

Presaging the harsh conditions was a trip out there that was tainted by inconvenience. I was driving and gave 16-year-old Dennis, an acquaintance from WIPS and the class, a ride too. Our route should have been straightforward and half freeway—I-70 as far as Rifle, then up a good blacktop road (Hwy 64) through Meeker and Dinosaur; maybe a 6 or 7 hour trip, so we left from Dennis' house in the late morning, figuring it should be no problem getting to the campsite before dark. Well, any of you who are familiar with I-70 through Colorado know the stretch through Glenwood Canyon: winding and very scenic, a limestone gorge along the Colorado River that was the last stretch of the originally-planned interstate highway system to be finished (only in the last decade), because of the difficulty of building the interstate through a high, narrow canyon hundreds of feet deep, with sheer rock walls. And you know that roads through the mountains are few and far between. Well, wouldn't you know it, but just as we reached the mouth of the canyon, we got stopped dead in a humongous traffic jam, with cops everywhere. We had no clue what was going on, but it was obvious that we were going nowhere fast. We weren't too far past Wolcott, where highway 131 takes off to the north towards Steamboat Springs, and our only chance of going a different way around, albeit with a three-hour detour—so after sitting at a dead stop for maybe 15 minutes and deciding that staying on the freeway was not an option, we pulled a U-turn over the median (4-wheel drive is nice) and committed ourselves to taking

the long way around. There was no way of knowing when the freeway might open up again, or how many miles ahead the problem was,
but I'd rather be driving and know I'm making progress than just
sitting in a traffic jam, and Dennis agreed.

So of course it was twilight and getting darker by the minute
when we finally regained our route out of Dinosaur and were trying
to follow Kirk's directions to the campsite. Fossil outcrops rarely
have the good manners to position themselves right next to good
blacktop roads, and neither did the campsite where we were to hook
up with the rest of the crew.
Even when the roads are
paved, these county byways
in the middle of nowhere are
sparsely marked and difficult
to find in the dark, especially
when your directions say
"pass the silo on the left and
turn onto the next road you
see." But we had a cooler full
of near beer and plenty of
granola bars with us, so we
did ok, and only had to back-
track a couple of times before
we found our way and pulled
into the campsite. It wasn't
really a campground, just a
grass-and-sagebrush flat with
a lot of cottonwoods by the

Macginitiea, an extinct sycamore, now displayed
at the Denver Museum.

White River, so at least we didn't have to worry about all the sites
being taken by 10:00 at night—you just set up your tent in an open
spot and hit the sack. By talking to some of the other participants,
we found out what the mishap on the freeway had been: an oil tanker
had overturned in the westbound lanes, spilling its slippery cargo all
over the pavement. One of the other members of our group had

come by just after it happened, and said that cars were hitting the slick and sliding off the road everywhere, and that they had barely made it through. So the cops had closed the westbound half of the freeway until a scrubbing operation could clean up the mess. We were glad we had decided to go around, because that sounded like it probably would have taken the better part of the day.

Parvileguminophyllum found by Darlene Emry.

Refreshed the next morning, we all caravaned out to the dig site. It at least had enough manners to be a short walk from a good gravel road, although we did have to climb a fair ways up a steep hill to get to the proper stratum. The rocks we were excavating were the Parachute Member of the Eocene Green River Formation. The Green River is a series of lacustrine shales (including some of the best oil shale deposits in the country) that were deposited in three large lakes that existed in the area where Wyoming, Colorado, and Utah come together back in the Eocene: lakes Uinta, Gosiute, and Fossil. The deposits are world-renowned for their beautifully-detailed fossils which are preserved in the very fine-grained greyish-brown shales. Depending on what layer of the formation you are in, you may find insects, leaves, fish, or even other vertebrates—the earliest known bat, *Icaronycteris*, comes from the Green River. In fact, two other trips I've described in this book, the one to Douglas Pass and the fish dig in southwestern Wyoming, were both in the Green River Formation. The best fish come from the remnants of Lake Gosiute and Fossil Lake, whereas the leaves we were excavating near Bonanza represent sediments deposited during a maximum high stand of Lake Uinta.

Offhand, one would think that leaves found in lake deposits would have come from trees growing quite near the shore. However, some of Kirk's work has indicated that the Bonanza locality represents a flora that includes species that grew as much as 40 km from the site where they were deposited, complicating efforts to reconstruct the paleoecology of the site. The bottom waters of these lakes frequently went through anoxic episodes (spells during which they were low or lacking in oxygen), and this probably accounts for the beautiful preservation of the leaves, because decay organisms were few or absent.

While the Parachute Member of the Green River has been well-studied in some areas, Kirk's aim for our dig was to document the abundance of common species and try to reconstruct the paleoenvironment and lakeshore vegetational community. This means that we weren't just collecting museum-quality specimens, but pretty much every leaf we found that was well enough preserved to be identifiable as to genus. The dig site was positioned on a natural outcrop near the top of a moderate-sized hill, where the slope of the hillside was concurrent with the dip of the strata, so that there wasn't any overburden to remove or pit to dig. You just found a convenient spot and sat down on the bare rock. With your rock hammer, you would whack away at a layer until flagstone-like pieces separated along the bedding planes and became free. Depending on the thickness of these flagstones, you might crack them into thinner pieces or not, and then examine both surfaces for carbonized imprints of fossil leaves. Everything that looked promising was set aside, and every now and then when you needed a stretch, you'd mosey on over with your finds to a picnic table Kirk had set up, and he'd look them over and

Rhus, a sumac, also now on display at the Denver Museum.

decide which ones were keepers. The keepers were carefully numbered with a magic marker and wrapped in toilet paper to protect them, and then placed in boxes in the back of Kirk's truck. The fragmentary, unidentifiable ones you could either keep yourself or chuck down the hill.

Physically, the work was not difficult, as the shale split fairly easily, but also had enough coherence that you weren't working with playing-card-thin, fragile pieces like we had been at Florissant. But the weather was not overly kind to us, and that dampened my spirits a little. The low 40s would not have been too bad except for the incessant, biting wind, and of course nothing but sagebrush to break it. Even with my goatskin gloves on, my fingers would be so stiff after an hour that I could hardly handle the shale, and I'd have to take a break and go warm up in the truck before going back to work.

Populus, a poplar or aspen.

But by having twenty or so of us work the outcrop for three days, what a bonanza of fossil leaves we unearthed! Most of them were from angiosperm trees, and some specimens were absolutely perfect. The most common types we found were *Parvileguminophyllum* (a compound-leafed tree of uncertain affinity resembling a legume, or member of the Leguminosae/pea-bean-acacia family); *Cedrelospermum* (Ulmaceae, elm family); *Macginitiea* (Platanaceae, sycamore family); *Rhus* (Anacardiaceae, cashew family, including sumacs); *Allophyllus* (Sapindaceae, soapberry family); and *Cardiospermum* (another soapberry). Altogether, the flora and other evidence point to a moist, subtropical or warm-temperate environment, consistent with what is known about the paleoclimate of this part of North America during the Eocene.

The site produced some of the best fossil leaves I've ever seen, and all in all, the dig was a fun and rewarding experience. The trip home was happily uneventful, and the leaves we found are now accesssioned into the collections of the Denver Museum, where several of the best ones are also on display. Currently, Kirk and several of his volunteers are working on a major monograph and catalog of the Bonanza flora which, when finished, will provide a definitive guide for fossil collectors and paleobotanical researchers alike.

In fact, I had so much fun that this summer my mom and I decided to take the kids back out there to hunt for our own leaves. The site is on BLM land, so you are allowed to collect non-commercial quantities of plant fossils, and this time, we didn't have to give the good ones to the museum. We didn't find anything that compares to the museum specimens shown on the previous pages, but we were pleased with our finds anyway. The spot we stopped at was on the same gravel road south of the river and east of the highway that the place we had gone to with Kirk had been, but it was a different hillside where most of the shale had already weathered loose. I like places like that where you can examine a lot of rock without doing a whole lot of digging. Allison found a big slab with three beautiful leaves in it, whose only flaw was that the slab was quite large compared with the size of the leaves and a

Allison collecting at Bonanza.

My *Macginitiea*.

My aspen.

bit difficult to carry back to the van. Two of the leaves were entire-margined, elongate forms that resemble willows, and the third was a toothed variety that could be a small elm leaf. I found a beautiful *Macginitiea* (the extinct sycamore) and an aspen leaf that shows absolutely perfect detail of the leaf veins. And this time, it was still just as windy, but the temperature was in the 70s so the collecting conditions were much pleasanter.

We didn't make near the haul that our first trip had, because there were only the four of us and we just collected for part of one afternoon before getting chased off the ridge by the approach of a large thunderstorm. But I think this is going to be one site that I'm going to want to return to again and again whenever I happen to be in the vicinity. It's certainly the best leaf site that I've ever been to, and it's only a day's drive from where I live, so I can see it becoming a regular haunt of mine when I'm out on the Western Slope.

18

Antelope Spring

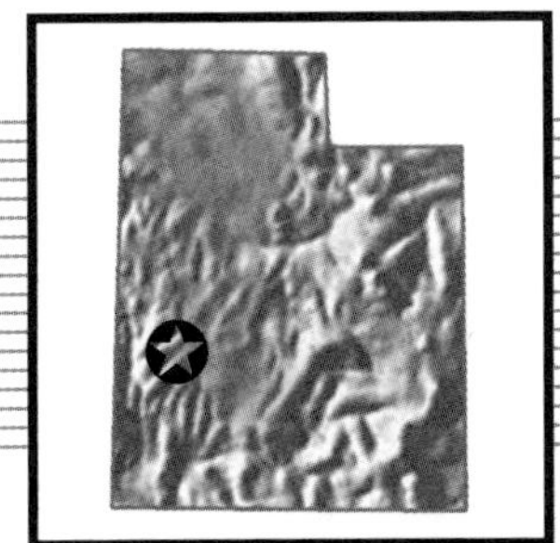

My first encounter with the trilobites of Antelope Spring, Utah was over twenty years ago, in 1978. We were living in Michigan at the time, and my parents, sister and I took a trip "out west" to see what there was to see. When we reached Utah, we headed south toward the basin and range country around Delta, because my mom had a sketchy description from a fossil guidebook indicating that trilobites could be found there. We headed west

View southeast from trilobite quarries with Sevier Dry Lake in the distance.

out of Delta on US 6-50 and took a turn to the right on an unmarked gravel road (not much more than a cowpath) snaking north through the Whirlwind Valley and into the House Range. Just about the time we thought we were lost, we encountered two guys in a truck who worked

Swasey Peak as seen from the trilobite digs. If you go too far up the Swasey Peak road, it becomes 4WD and difficult to turn around.

for the BLM. We stopped to chat, and one of them asked if we had anything for a headache. You know by now that headaches are my "Achilles' heel," and I always carry some kind of medication with me. We were glad to help him out. When they found out we were looking for trilobites, they said, of course, just continue up this road a couple more miles and into the hills. When you see the spring on the left, stop and start hunting.

We couldn't have imagined beforehand how prolific that hunting would be. We had barely parked the car when I found the first trilobite, a broken *Elrathia kingi* embedded in the dark grey rock of the Middle Cambrian Wheeler Shale. Even though it was a pretty poor specimen, I was ecstatic. Little did I know of the riches to come. In those days the whole area was pretty deserted, and you could roam the ridges and valleys to your heart's content. Where the arroyos were scoured down to bedrock, you found mostly impressions of trilobites in dark shale slabs. The slopes consisted of lighter material, weathered to clay, which produced whole black trilobites free of matrix. Most were *Elrathia*, ranging from practically micro-

Elrathia kingi, the most common trilobite from the Wheeler Amphitheater.

The shale on the beds of the arroyos yields trilobites still in matrix.

scopic to about an inch long. Many were missing the free cheeks, evidence of their having once been molts, rather than the remains of trilobites that had died. (As arthropods, trilobites molted their carapaces—shells made of fingernail-like material—periodically, so that they could grow.) The ones with a bigger glabella were *Asaphiscus wheeleri*. I quickly came to prefer the "free" trilobites over those embedded in the heavy, sharp slabs of matrix. Small inarticulate brachiopods sometimes were also found in the shale slabs, but were too delicate to weather free. Much less common than *Elrathia* and *Asaphiscus* were the little agnostids or "double-headed" trilobites. I think I only found three or four of them in the several days we were there, compared with bags and boxes full of the others.

The following year, I took Chris back there on our honeymoon. He didn't know much about fossils, but liked camping in the middle of nowhere. Some nights we camped near the trilobite beds, and when we wanted a bath, we drove back to Delta and north about 20 miles to the Little Sahara Recreation

Area, which has a nice campground with water at the foot of gigantic sand dunes. The noise of ATVs is disturbing during the day, but most days we were back at Antelope Spring hunting trilobites. The hills around the trilobite beds are covered in a sparse piñon-juniper forest, and the valley opens out to display a wonderful view of the Sevier (usually) Dry Lake in the valley to the southeast. More and more trilobites were added to my growing collection. They are so beautifully preserved that it is hard to pass one up, even if you already have a hundred like it. And there is always the allure of searching for "the big one." I did find an *Asaphiscus* on that trip that was nearly two inches in length, the biggest one I have ever found. (To this day, Chris claims he put it there for me to find.) After I returned home to my job at the GM Proving Grounds, I displayed some of the fossils I'd found on my desk for my co-workers to inspect. I remember Larry asking me how old they were, and I casually answered, "Oh, about 500 million years."

One of the locals. This was the smallest horned "toad" (really a lizard) I've ever seen.

He asked, didn't I mean 500 years? I was appalled to discover that he thought that fossils were formed when an animal died lying against a rock and was absorbed. How could an educated person be so ignorant of the planet around him?—I thought, but of course I didn't say it, and only explained in a few sentences how fossils were really formed.

The 1983 floods filled the Sevier Lake right up to the edge of the highway.

The biggest *Asaphiscus wheeleri* I ever found.

By the time we returned in 1983, we had moved to Colorado and it wasn't such a long drive (for us; my visiting parents still came from Michigan). The flooding Utah had experienced that year had filled the Sevier Lake. We walked across the salt flats down to the water's edge, and discovered that hundreds of fish had been washed downstream with the flood waters and killed by the salt water of the lake, littering the shoreline. Fortunately, because of all the salt, they didn't smell. By that year, we had our dog, Rooter, and my parents had their dog, Tilly. The dogs loved roaming the hills as much as we did, although they were more interested in sniffing cow pies than hunting trilobites.

Thirteen years and two childbirths passed before I was able to visit Antelope Spring again. I returned with my family in 1996 as a field trip leader for WIPS. Things sure have changed. Gone are the days when the trilobite digs are obscure and in the middle of no-

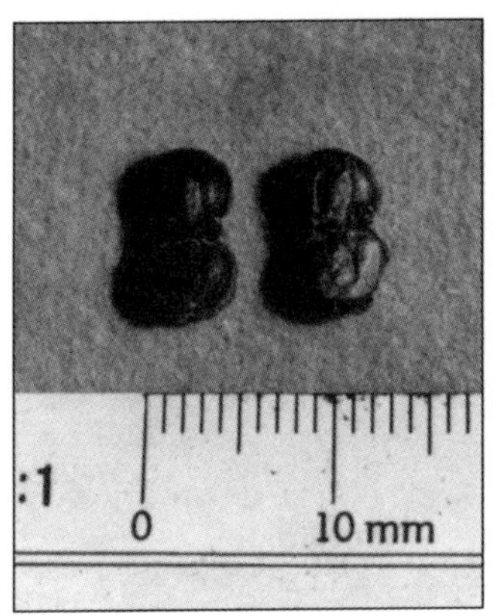

Agnostid trilobites, *Peronopsis interstricta.*

where. Sure, you still have to drive 40 miles from civilization on gravel roads, but they are graded now and marked with signs. A couple of guys in our party even made it in a motorhome. Twenty years ago there was only one commercial quarry (the area is a mosaic of state and BLM land, and the "mineral rights" to various parcels can be leased); now there are lots of "no trespass-ing" signs, and you have to be alert to avoid those areas. The biggest of the commer-cial operations is called "U-Dig Fossils," which charges you by the hour to look for trilobites in a huge quarry excavated by heavy equipment. The good thing about U-Dig is that they have posted big signs at every road junction directing you to the quarry, so you no longer have to worry about getting lost.

Of all the localities I have described in this book, Antelope Spring has to be my favorite, even if it

Mattie and Allison hunting for trilobites at Antelope Spring.

isn't as desolate as it used to be. And if you decide to try your luck at only one place mentioned here, this should be the one. It is not a place where you will roam or dig all day and come home with only a fragment or two. You do have to be careful nowadays to avoid the commercial quarries, but there is still plenty of BLM land you can roam freely on, and you should easily be able to return home with at least a small container of whole, beautifully-preserved trilobites. I have heard that other types of trilobites can be found in the younger sediments atop Marjum Pass, but I've never stopped to look. Maybe next time.

19

Dinosaurs from the Land
of the Midnight Sun

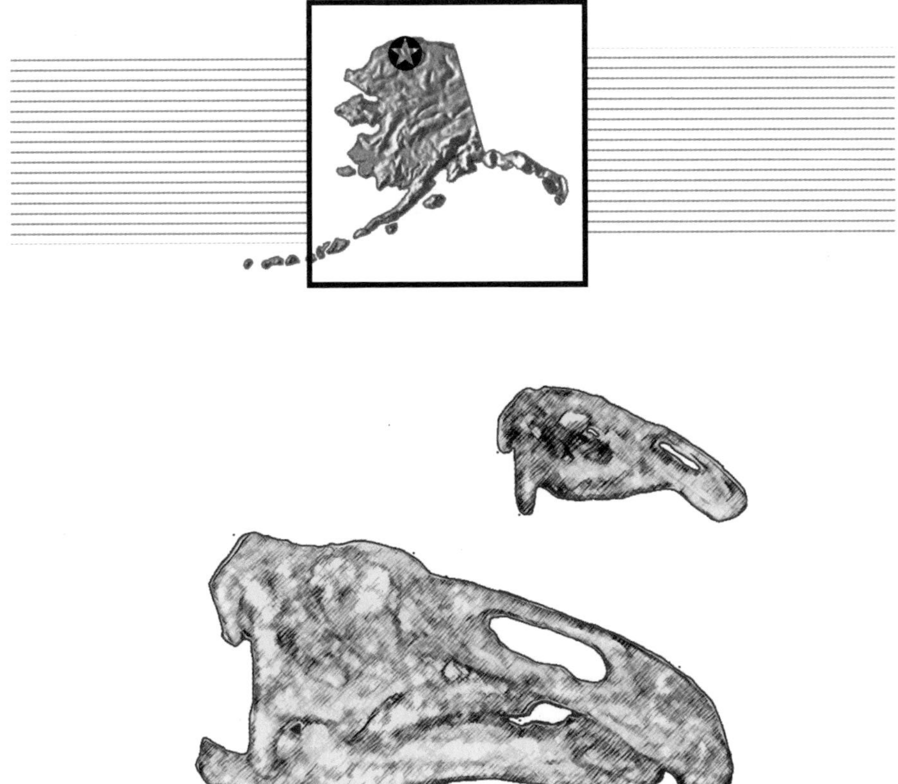

During the summer of 1996, I decided I needed a nice, long vacation from kids and housework, something I hadn't had since Karen's and my trip to South America two years before. I signed up for an expedition jointly sponsored by Dinamation and the University of Alaska—excavating a hadrosaur bonebed on the North Slope of Alaska, on the Colville River just 41 miles from the Arctic Ocean. This is the journal I kept during the trip.

In black and white, this photo doesn't do justice to the bright magenta color of this field of fireweed. Fireweed is one of the first herbs to colonize disturbed areas such as this forest which had burned.

Monday, July 15—flight to Fairbanks. Went to U of Alaska museum to see "Blue Babe," the Pleistocene bison which was found frozen in the ice a few years ago. It was really cool. It had been mounted over a core so it didn't look shrunken. I stayed up till after midnight, reading. It was still light when I went to bed.

Tuesday, July 16—Jonathan called and came to pick me up after breakfast. Took the last nice warm shower I'll probably have for a while. Picked up Bill and Margaret and went to UAF museum to meet the rest of the crew and Roland Gangloff, our leader. Had an orientation at the museum, then went to Gary's house to pack and pick up the boat. We'll be going in two university trucks and pulling a 16' aluminum boat, which will also serve as a trailer to hold food and gear. We hit the road about 4 pm: the Dalton Highway, also known as the "haul road," built to service the Alaska Pipeline. Got our first glimpse of the pipeline, which will be our constant compan-

I'm standing next to the Arctic Circle sign, latitude 66°33'.

ion as far as Prudhoe Bay. It's about 75°F—much warmer than I expected; I need the clothes I wore from Denver. Saw an amazing field of fireweed where a forest fire had been sometime ago. We camped 5 miles north of the Yukon River at a site with a water well. Enjoyed a campfire, baked potato, and a cool beer for dinner.

Wednesday, July 17—it didn't get dark at all last night even though we're still south of the Arctic Circle. The sun did set, but it stayed like twilight. Woke to a crystal blue morning and 70° temperature at 8 am. Washed my hair at the well—cold but good. Hit the road shortly after 9. Got to the Arctic Circle around noon, took pictures of the sign and had lunch. Stopped in Coldfoot to buy and mail a few postcards. By the time we reached Atigun Pass in the Brooks Range it had cooled off a lot. I put on the Alpaca wool sweater I got in Peru. We saw a dozen white Dall sheep crossing the road. We collected snow for the coolers at a small residual snowfield near the pass. We've left all the spruces of the boreal forest behind, and now it's all tundra punctuated with dwarf willows. We're only a mile or so high, but it looks a lot like Colorado at 12,000 ft.

Dall sheep, Atigun Pass.

We make camp at Galbraith Lake. Now I believe the stories about Alaska's mosquitoes—not bigger than Michigan's, but vastly more numerous and voracious. Ed wants to wash his hair again in the morning, as I do. We scout out a small stream after dinner, and plan to come back in the morning to wash up. By 8:30 we've all retreated to our tents to read and sleep, even though the sun's still high in the sky and it looks like Boulder at 6 pm. I'm sure glad I brought this bug-net suit!

Thursday, July 18—hit the road about 9:30 this morning. Spotted 29 Dall sheep and a couple of sikriks (arctic ground squirrels). Made camp at Happy Valley. The mosquitoes are phenomenal—they call them the Alaska state bird, with good reason. This bug suit has already paid for itself. Even so I've retreated to my tent to escape them. We are all getting to know each other, which makes the expedition more fun. Believe it or not, I'm too hot—it's 70° and I only brought cold-weather clothes. We saw a caribou at close range, and even chased him a ways down the road. The tundra is beautiful with sphagnum moss, horsetails, cotton grass, lupine, and dwarf willows.

Friday, July 19—made it to Deadhorse/Prudhoe Bay this morning. Not a very scenic town. The plane only holds the pilot and 3 passengers, so will have to make multiple runs. Ed is a real trooper but he's not having a good time. He's covered with mosquito bites and likes to be clean. He's a librarian from St. Louis who has never been camping before. He's sure got spunk signing up for a trip like this. I know he wishes he could back out. Judy and I have both wondered what we've gotten ourselves into, but at least we're veteran campers. Did get a hot shower at the Prudhoe Bay hotel—felt great.

6 pm: got out on the third flight with Margaret and Barbara. It was very interesting flying over the tundra. Saw a herd of caribou from the plane. Beautiful animals. From the air you can really see the geometrically-patterned ground. I've forgotten just how it forms but I think it has something to do with frost heaving. Saw a pingo from the air, too. It's a little hill with an ice lens at the core which provides the only relief on this arctic coastal plain. The tundra is punctuated by hundreds of small lakes. You can still see vehicle tracks 30 years old left over from the early days of oil exploration. The tundra is truly a fragile environment, just like our alpine tundra in Colorado.

The flight to camp on Poverty Bar is about an hour. It's drizzling and we fly below the 800 ft. cloud ceiling. The pilot lands us on a gravel bar by the Colville River, where the first two crews have already set up their tents and the "Hansen structures," tentlike quonset huts which will be our meal quarters. We even have a camp manager/cook—such luxury—who has made special provisions for my being a vegetarian. I have to set my tent up in the mud and drizzle and it appears that it's not 100% waterproof on the roof. I can only pray that the rain fly will keep any more water off the roof other than that which got on it while I was setting it up, 'cause I'm stuck here now, for better or for worse.

Did I mention that we've had a campfire every night? Nice little perk. Even here on the tundra there's wood to scavenge from the dwarf willows, and small logs that wash downstream from the Brooks Range. Camp is carpeted with some type of fuchsia flower, the color of fireweed but shorter and with bigger blossoms. I hope someone will be able to tell me what it is. After setting up my tent, I throw my

Camp on Poverty Bar. My tent is the pyramid-style one in the center.

Looking north (downstream) along the Colville River. The quarries are off on
the bank at left.

gear inside to get it out of the rain. Already I'm beginning to get a
cover of grey mud that I suspect will be ubiquitous before this is
over. I sit and listen to the rain on the roof of my tent. It's not too
cold, maybe 60°, and so far I'm comfy. The mosquitoes are here but
not in such force. I am having fun, so far, at least. I think about poor
Ed, who's due on the next flight. I think he secretly hopes the weather
will be bad enough that he and Roland will be stuck in Deadhorse for
a night at the hotel. You don't have to worry about losing your light
for flying—there's no darkness here in summer. It makes it a bit hard
to sleep so I'm glad I brought an eyeshade.

We're really developing some camaraderie as we get to know
each other. Half of us signed up through Dinamation and the other
half are Alaskans who are taking this as a course through the university. The Alaskans are Barb, a retired librarian; Erica, a college geology student; Matt, an 8th grader; George, an Eskimo whaler from
Barrow; Ron, an Eskimo student from Barrow; Gary, the camp cook,

a history teacher who currently works at the museum; and Roland and Dave, the two professors. Our Dinamation cohort consists of Ed, whom I've mentioned; Judy, who lives in Portland; Bill and Margaret, who are from San José; and Jonathan and me, who are both from Colorado.

Dinner is tortillas with cheese because all of the food hasn't arrived yet. It's still raining but Bill gets the fire going by dousing it with white gas.

Saturday, July 20—I woke up at 3 am to find my tent coming half down on me because the wind had pulled out a stake. I had to hammer them all in and put stones on top of them. The rain is abating somewhat. This morning Barb's thermometer said 44°. Breakfast is leftover noodles and black coffee. I spend most of the morning in my tent, reading. Ed and Roland fly in with the last of the gear and spend the day organizing the camp. The rain has stopped by late afternoon and it gets almost clear, and is up to 61°. Even so, there are mosquitoes and mud everywhere. The mud cakes to my boots and gets in my tent. I wish I'd bought a new dome tent instead of using my old EMS pyramid, but it's too late now. I keep thinking, what have I gotten myself into? I hate the mud and the zipper on my tent's mosquito netting has broken.

Sunday, July 21—I slept much better last night and the rain has cleared. This morning it's about 55° and partly cloudy and dry. I hope it holds up. Some of the mud is drying up, which I'm delighted to see. Breakfast is tang, coffee and pop tarts. The mosquitoes are already out in full force despite a little breeze.

Great day at the outcrop. We are digging in the Prince Creek Formation of the Colville Group, Cretaceous, 71 million years old. Some people walked up there (about a mile) while the rest of us took the boat. Dave and Jonathan drove the boat from Deadhorse, where we left the vehicles, to Poverty Bar, where we are camped. There's been a lot of slumping in the past two years and quarries need to be dug out to reach the Liscomb Bone Bed, our objective. Some of us sort through talus while others dig out the quarries. In the talus I find two vertebral centra from the tail of a hadrosaur and two teeth of an albertosaur (one dentary and one premaxillary), complete with serrations, the find of the day. I decide to walk back to camp with Erica and Matt, picking up a load of wood on the way. In addition to the other sticks, you can sometimes find Pleistocene wood weathering out of the Gubik Formation which unconformably overlies the Prince Creek. The Pleistocene also produces beautiful clam and snail shells which wash down onto the grey sands of the beach. The walk is long because we have to slog through ankle-deep muck the whole way, which threatens to pull off our boots. I'll take the boat after this.

The British film crew has arrived and they film us having dinner around the campfire. It's getting much colder and has begun to

Bill, Margaret, Matt, Gary, and Ron opening a quarry.

sprinkle so I retire to my tent to write and read. I've got two broken zippers on my tent now and will definitely get a new one before embarking on any more expeditions. It's also getting windy and I sure hope my tent stakes hold.

Monday, July 22—woke up this morning to sunshine, calm conditions and no bugs. Last night's rain was short-lived and I had no more problems with the tent. Took the boat out after a breakfast of scrambled eggs and coffee. Gary really takes care of us. Spent the morning opening a quarry—shoveling off overburden of slumped "Gubik goo"—only to reach permafrost in one direction and have the bone layer pinch out the other way. Gary brought our lunch of pasta salad and crackers around noon. Spent the afternoon trying to find the bone layer to the south, then prospecting up the beach where I found some very nice petrified wood—coniferous, Roland says, either Araucariaceous or *Parataxodium*—coming from the same beds as our dinosaurs. Later I moved on down to Ed and Judy's quarry, where they had found the bone layer and were setting rebar for reference grids. I think I'll dig with them tomorrow. We had to wait a long time for the boat because Dave took the TV crew upriver to photograph ice wedges. Finally around 6:30 Jonathan came for us in the other boat. I may walk back tomorrow, after all. The weather held great all day, maybe 60° and sunny, but the wind shifted to the north around 4 pm and it got a lot colder: Arctic Ocean winds.

The peregrine falcons nesting on the tundra above the digs screeched at us all day. Jonathan and Erica found their nest with two downy chicks in it. There are rough-legged hawks nesting across the slough from our camp on Poverty Bar. We have a telescope trained on them and occasionally see the chicks being fed.

Tuesday, July 23—another fun day on the Colville River, thanks to a smiling weather god. It was cool and a few sprinkles this morning. I've gone back to just having coffee—these camp breakfasts are just too much. We opened our quarry pit this morning—the grid was laid yesterday—and worked down through the clay to the bone layer, which is a black, barely lithified shale. Lunch was a feast of macaroni soup plus peanut butter on whole wheat and apples. After lunch Judy uncovered what she thought was just a rock—rare in these fine sediments—but which I knew was a partial caudal vertebral centrum, about 2" diameter. We've been told to expect many of these flattened, spool-shaped bones from the tails of juvenile hadrosaurs. She also found a rib fragment and several ossified tendons. We measured the x-y-z coordinates of these and drew them in on the map. Ed and I worked the back side of the pit all day but didn't find any fossils. The pit is now about 8" deep. We covered the quarry and rode the boat back about 6 pm. There was no wind today and the sun came out in midafternoon, so the mosquitoes were really intense, but fortunately, I was protected by my trusty bugnet suit. Back in camp, as I write in my journal, I enjoy a cool near beer, thanks to Stu, our pilot, who picked some up for me in Deadhorse yesterday. It really hits the spot after a

Hadrosaur bones found in talus. Upper left: end of longbone; upper right: vertebral centrum; lower right: unidentified.

long day at the digs. I find myself getting really thirsty up here and wonder if the atmosphere is dry despite all the water on the tundra. Barb says they only get about 15" of precipitation a year, the same as Denver, but with the cold up here and the incredible flatness of the tundra, underlain by permafrost, meltwater stays around.

Wednesday, July 24—another day of great weather, sunny and in the 60s. We got down to the business of really excavating the bonebed today. Judy found a metatarsal II from a juvenile hadrosaur, Ed found a cervical centrum and another metacarpal. I found several bone fragments, but that's all. Late in the day Ed also found a carnivore tooth, serrated, looks much like my albertosaur. We mapped them all in on our quarry sheets. There was a breeze to keep the bugs down. One of the peregrine chicks fell out of the nest and Jonathan climbed up the bluff and put it back. He thinks the parents are spending all their time squawking at us and not feeding the chicks, so he gave them some fish, and they ate it. The film crew finished up their photography today. They took a few shots of us digging but spent most of their time interviewing Roland. It's pretty interesting to see how they work.

All three of the zippers on my tent have now broken or gotten stuck. I definitely need a new tent. Dave got them going again and tightened them up, then I smeared them with Bag Balm to lubricate them because that was all I had. It's beginning to rain again but is still warm.

Thursday, July 25—I walked with Jonathan to the site this morning because I was doing dishes and missed the boat. We stayed off the mud as much as we could, but had to cross the flats where we saw tracks of sandhill cranes, caribou, and a bear. Today started out cool and sprinkly but quickly warmed up when the sun came out. It must have been close to 80° this afternoon. I found one complete phalanx 2 from digit II of a juvenile hadrosaur foot. The rest were just rib fragments and more ossified tendons. Jonathan found a second metatarsal—must be from a subadult because it's much bigger. Roland thinks there was a crèche of young which died while crossing a river and then were redeposited in a crevasse splay. The film crew left today. I really wish I got the Learning Channel so that I could see the episode of PaleoWorld for which they were filming; maybe I can find someone back home who gets it and would tape it for me. Back to camp and I took a shower at the sun shower and I feel so much better. After dinner it was so clear that we could see the Brooks Range, 150 miles to the south. As long as the weather holds, I'm having a great time. Apprehensive as I was about this trip, it's turning out that I was very well-prepared, and that makes all the difference in the world. The things that worried me most were insects and getting cold; but my bug suit keeps the mosquitoes off, we haven't encountered any black flies, the weather has been better than I expected, and I have plenty of warm, dry clothes. So none of my worries has come to pass, the people are wonderful, and the bones are plentiful; this has to be one of my best vacations ever.

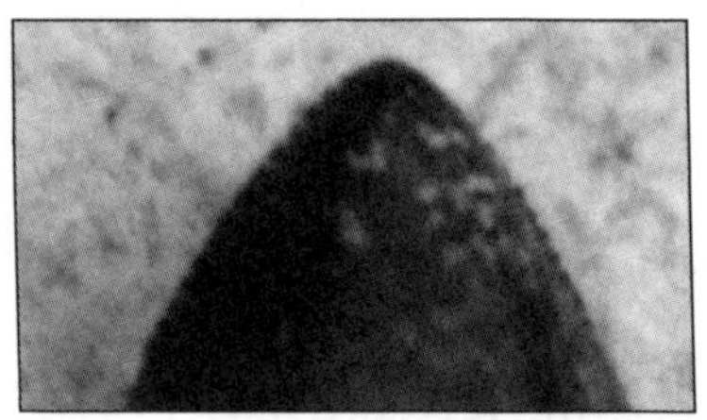

Albertosaurus tooth (above) showing serrations (below).

Friday, July 26—another day that started out cool and sprinkly, then warmed and cleared, but now that it's dinnertime it's raining again and blowing quite a bit. I hope my tent stands up to the wind. We excavated a few more centimeters into the bonebed today and are now coming to the layers with the big bones. It appears that there was some measure of stratification in that the biggest bones are at the bottom of the layer and are overlain by a fining-upward sequence resulting in mostly teeth, bone fragments, and ossified tendons near the top. I found two caudal centra and Ed and Judy each found a humerus, all from *Edmontosaurus*. Ed has adjusted well and I think he's having a good time now, at least when it cools off and the bugs abate. We may not be able to salvage Ed's humerus, because it's very punky and fractured from being in the active freeze-thaw layer, but the quarry data on location and dip will still be valuable information. Judy's humerus is in better shape, being located further into the hill, closer to the permafrost. We soaked it with Vinac and left it to dry till tomorrow, although how much drying it can do under visqueen in the rain is debatable. The rock is wet with percolation as the permafrost thaws when our excavations bring it closer to the surface. At least our quarry pit isn't crosshatched with ice lenses like Gary's is, though we have some big fractures where they used to be. Jonathan recovered a nice astragalus today and Matt found two *Albertosaurus* teeth, one pristine and with a partial root. All the remains we are finding are from juvenile *Edmontosaurus* except for the shed carnivore teeth, possibly evidence of scavenging.

Dave went downriver and shot a caribou which he butchered and hauled back to camp in the boat. I walked back a) to try to get back before the approaching storm and b) to avoid riding in a bloody boat. Hopefully the rain will wash it clean. Dave also caught five arctic char (a type of salmon) which is what they're having for dinner tonight. Barb's not a vegetarian but she's allergic to fish and doesn't like caribou, so the two of us will partake of lighter fare. It's a wonder they can keep the fire going in this rain to cook the char. There's even thunder and lightning, phenomena which are rare on the North Slope.

We saw several sikriks today. They're cute, bold little critters, with long fluffy tails and spotted beige backs; otherwise, they look like slender prairie dogs. They run down the bluffs and across the beach and sniff for food and knock small rocks down on us. Roland and Dave don't like them much, but I think they're charming.

Saturday, July 27—the rain stopped during the night but the clouds remained, and it was 42° this morning and I don't think it warmed up all day. At least it didn't rain. I need all my warm clothes now. I had on my leotard, my long underwear, my fleece pullover, my down vest, and my Gore-tex parka—and was just comfortable. I also needed my stocking cap. Rounding out my quarry attire were sweats under my Gore-tex pants, and wool socks under my rubber boots. We need the rubber boots for wading to/from the boats or for crossing the mud flats if walking. I'm glad I invested in the Gore-tex: it's wind- as well as rainproof, and perfectly suited to this weather. The upside to the cold is that the mosquitoes weren't out. I had three cups of coffee for breakfast. This is camp coffee: muddy Colville River water, boiled for ten minutes to kill any organisms, then reboiled with some ground coffee thrown in. You strain it through a sieve into your cup, and it's still a little crunchy at the bottom. It's nice and strong and thanks to Gary and Stu, we still have milk—most people use it on cereal but I dump my ration in my coffee. Having milk for my coffee is heavenly. Ed is still having trouble in that he can't stand to drink the muddy water, so is rationing juice drinks he brought from home. I don't like it plain, but Gary covers the mud up with Crystal Light each day, which makes it quite palatable.

I excavated the north quadrant of our quarry pit today and found two more caudal centra. You can tell the caudal centra from the dorsals and the cervicals because the caudals are amphicoelous whereas the others are procoelous. Judy worked on a radius and Ed worked on his humerus. The next quarry pit over from ours is that of Erica, Barb, and George; they have been working on excavating a very nice tibia.

Gary brought hot tortellinis for lunch, which really hit the spot on a cold day. At least it hasn't been windy. I leave the recording to Ed and Judy because my fingers are too stiff from the cold for me to write legibly, despite my goatskin work gloves. My fingers are cold, as are my toes, but not terribly uncomfortably so. I'm actually quite comfy under the circumstances. Erica and Jonathan are talking about building a "river sauna" tonight—some sort of fire and hot rock bake followed by a dunk in the muddy Colville slough. I think I'll pass. I might be up for that sort of thing if the air temperature were warmer, but not when it's 42°.

After lunch I found a beautifully preserved neural arch, complete with neural spine and zygapophyses, from a caudal vertebra, lying just above a nearly complete scapula. The coracoid end has turned to punky powder but the rest of the bone should be recoverable. I can't go far enough to jacket either one until Judy's radius and humerus are removed. Ed covers the radius with toilet paper, then dampens it with water and a paintbrush; the next step is to plaster it with presoaked surgical bandages. The paper layer will allow the jacket to cleanly separate from the bone in the lab. I've never used preplastered bandages before, only burlap strips dipped in plaster of Paris. The bandages are easier to use but are more expensive. They were a donation to the expedition by a veterinarian over a year ago, and apparently, some of them have gone stale: they absorbed some moisture through the packaging and the plaster has partially set up, so they're crunchy and grainy when soaked rather than muddy and fluid. Bummer, because the extra plaster got accidentally left behind in Fairbanks. We can only hope we have enough good ones to do the job. In this cold, the plaster must be allowed to set up overnight before the blocks can be pedestaled and overturned in preparation for completing the jackets. Around 4 pm Jonathan brings the big boat back from camp with the makings of an afternoon tea. Dave has taken the little boat downriver to the Eskimo hamlet of Nuiqsut to visit friends. A fire of willow branches and peat clods on the beach serves to heat the water, and we have hot chocolate and herbal teas—it's really wonderful. I can't get over the little luxuries we have in this remote spot. Tonight they're going to smoke the caribou ribs. It's threatening to rain again as I crawl into my tent to write in my journal.

Edmontosaurus metatarsal (foot bone).

Ed and Judy taking quarry measurements. Note Ed's mosquito netting.

Sunday, July 28—it was cool today, with high clouds, not cold like yesterday but cool enough to keep the bugs down—53° at 6 pm which is still midday here. We finished jacketing the radius and neural arch this morning and removed them from the quarry so we could do some more work on the humerus and scapula. Ed found a hadrosaur tooth and a nice, nearly complete rib near the middle of the quarry, but we spent most of the time working on bones we'd previously discovered. Our meter-square pit is deep enough that you must be in it to work now, you can't just lean over the edge, and its confines are only sufficient for one or two of us to work at a time now, which slows us down. The work is slow anyway, jacketing first one side, then allowing it to set up, pedestaling, flipping, and jacketing the remaining side. Fun, though. Neither Ed nor Judy had done any jacketing before so they were excited.

Barb, Erica, and George have reached the bottom of the bone layer in their quarry. Then it's just clay again. Their pit is only 2 meters northeast of ours, and they've found the bone layer to be only half a meter thick; we must be nearing the bottom of it, then. After another tea at the "Peregrine Quarry" (that of Bill, Margaret, Jonathan, Matt, and Ron; so named for the birds nesting above) we jacket the top sides of the humerus and scapula, then close up the quarry for the day. The plaster is setting up a little faster with the somewhat warmer temperatures today. It's been fine fun, dry and pleasant, though the mosquitoes are beginning to come out. The caribou ribs are smoked and they plan another meal of meat and fresh fish.

I washed my hair in the river before dinner and it felt good. The water was very cold, but I've done this before when I was camping in the mountains—sometimes in glacial meltwater—so it's not a big deal. Being able to keep clean contributes immensely to my comfort. Gary has made pasta al fiorno and carrot stix to supplement the meat. During dinner the wind shifts to the north and it begins to get cold. They are planning another river sauna for tonight, but again, I'd rather skip it with this cold air.

Monday, July 29—a cold night, but I put on my down booties and vest and snuggled into my down sleeping bag and I was toasty warm. I didn't want to get up this morning. Jonathan, Erica, Bill, and Margaret had another sauna last night and then went up on the tundra to watch the first sunset. The sun sets behind the bluffs now for an hour or so but it still doesn't get dark at night. I'm pretty used to it now and sleeping well.

It was 42° again at breakfast, but then the wind swung around to the north, coming in off the Arctic Ocean (only 41 miles away by river), and it grew quite perceptibly colder—though whether it was actually a temperature drop, or just a wind chill, I couldn't tell. It alternated between spitting on us and sprinkling all day, never really reaching a drizzle but enough to keep things wet. This was to be our last day of digging, and between that and the weather, I couldn't really get into it today. Neither could Ed, Erica, or Judy; we flipped some jackets and mapped some small bones, but the pace was distinctly slower than it has been. Judy found a nice fourth phalanx and I found another rib underneath where my scapula was removed, and I also found another hadrosaur tooth. The teeth are found in isolation here, not as part of the large dental battery they formed in life.

The lunch boat brought the supplies for tea, complete with blueberry muffin mix to be cooked in a small rusted wood stove/oven that Bill found on the tundra and hauled down to the quarry. We buttoned up our pit around 3 pm due to the rain, and headed down the beach to Peregrine Quarry. I was tired of the cold and rain and decided to skip the tea and walk back to camp early, rather than spend another two or three hours outside until the boat came. No one else wanted to come so I walked alone, and it was a pleasant walk since I could go at my own pace. The mud flats between the quarries and Poverty Bar are crisscrossed with ever so many footprints—caribou, heron, sandhill crane, bear, and sikrik, as well as others I couldn't identify. In places they are covered with mats of the prostrate, filamentous stems of creeping horsetails *(Equisetum)*, which color the mud flats green. We saw a musk ox today on the far side of the river, but it was too far away to see very well and no one had binoculars with them. Once we got back to camp, Jonathan got out his spotting scope and then we could see the musk ox very well. Magnificent animal.

As a celebration of our last day of quarrying, Gary smoked and roasted two turkeys and prepared a full "Thanksgiving" dinner—stuffing, green beans, cranberry sauce, and cheesecake. It was wonderful. I had noodles instead of turkey, but partook of all the trimmings. I sure didn't expect such great food when I signed up for this expedition.

Tuesday, July 30—the day broke cool again, 41° but dry. After a leisurely breakfast of French toast and coffee, Roland called a meeting to wrap up. The ribs and phalanx we left in our quarry will remain for the second crew to remove. This morning we are to stay in camp, finish the labeling of bones and our paperwork, and clean up the camp. Gary and Ron are staying here to provide continuity between us and the next crew, but the rest of us will be leaving tomorrow. Tonight we'll take a picnic to Kikiakorak up the river a little ways. With some time to kill in camp, Ed, Judy, and I climbed the bluff and took a walk on the tundra. We saw birds, sikriks, and the ubiquitous trenches that form the patterned ground. There were some kind of dwarf bushes prolific with red berries, many short herbs, and some beautiful clubmosses. From the top of the bluffs we could see far to the east, where the land is lower, over the marshy tundra bedecked with dozens of shining jeweled lakes. As we came back to camp, dragging willow limbs, the clouds came over the sun and it began to sprinkle.

I missed the boats to the picnic and caught a ride a little later with Just, the friend of Dave's who lives in Nuiqsut with his wife Inge-lise. They're both teachers and seem really nice. I appreciated Just giving me a personal ride in his boat so that I wouldn't have to

Jonathan and me in Nuiqsut.

forego the picnic. Apparently all I missed was a very cold boat ride in the rain, so it was all for the best. I had a late dinner topped off with a few roasted marshmallows, and Roland broke out the Scotch which the film crew had given us. We had a wonderful big campfire, a fitting end to a delightful expedition.

They finally convinced me to try the sauna. I still wasn't keen on jumping in the river so we took the sun shower into the teepee for me to rinse off with. Bill, Margaret, Erica, and Jonathan built a big fire and heated up some rocks in it. Then they shoveled them into a stainless steel salad bowl and brought it inside the teepee, and spritzed water on the rocks. It made a wonderful steam bath and we all sweated and lathered up our hair. Then the four of them ran out and jumped in the slough while I stayed inside and showered off. It was really fun and nice to be clean. Then we all warmed up by the fire and sipped cognac while we dressed. I'm glad I finally did it, even if I didn't jump in the river.

Wednesday, July 31—another cold, cloudy, but dry day. Just and Inge-lise are going back downriver and have invited us to spend the day in Nuiqsut. Jonathan, George, Ed, and I take them up on their hospitality; the others aren't as interested in seeing an Eskimo village, so they opt to stay on the bar in hopes of getting an earlier flight out and a hot shower in Deadhorse.

We stopped at a pingo on the way. Pingoes are small hills formed when the waterlogged soil of an old lakebed freezes, forming an ice lens which humps up the tundra soil. It's pretty interesting from a geological standpoint, and is also pretty, sporting as it does a mantle of yellow Indian paintbrushes.

Nuiqsut is a rather drab little settlement, population 350, but interesting nevertheless. The houses are all on pilings to keep them from melting the permafrost and sinking into the resulting muck. At the general store Jonathan inquired about purchasing some baleen, and we were directed to the red house, where an old Eskimo woman, the wife of a whaling captain, sold him an 8' plate for $50. Just and Inge-lise's house is very comfortable and we are all glad we came, both to see the village and to enjoy the comforts of civilization after all this time (warmth and a hot shower). The temperature has been dropping all day, and after dinner it's 38° with a stiff wind and a light sprinkling—I'm glad not to be spending the night on the bar, and feel for those who are still there. Dave and Jonathan talked to the pilot, Stu, on the phone, and found out he just got into Deadhorse today from a maintenance trip to Fairbanks; he got back too late to pick up the crew on the bar before tomorrow, delaying us a day.

Just showed us their ice cellar, which is a cold storage room dug into the permafrost. We climbed down a ladder past ice stalactites into a small cavelike chamber lined with beautiful, huge ice crystals on the ceiling and walls. It's a beautiful but foul-smelling hole with a floor made of frozen blood. They store fish and whale meat down there. I find the blood and chunks of meat a macabre sight, but am glad to have had the chance to see this unique storage system.

Thursday, August 1—the water has risen above the landing strip on Poverty Bar, so Stu can't pick the others up. They all pile into the remaining boat and come down to Nuiqsut, which has an airfield. We all fly to Deadhorse where we have dinner and then hit the road about 8:30 pm.

Friday, August 2—it was raining when we got to Coldfoot around 3 am,

Ed climbing the ladder down into Just and Inge-lise's ice cellar.

and no one wanted to set up our tents, so we drove straight through to Fairbanks, getting in around noon. Erica and I had all sorts of deep philosophical conversations as I made sure she stayed awake to drive, so I didn't get any sleep either. Barb invited us all to her cabin at Chena Hot Springs, so we (all but Judy and Roland) went up there this afternoon. Soaking in the hot springs was wonderful, then we all went back to the cabin to drink beer, play hearts, and talk. Bill and I stayed up until almost midnight, and after 40 sleepless hours, I was asleep as soon as my head hit the pillow.

Saturday, August 3—more soaking in the hot springs before packing up to go back to Fairbanks. Gary's mother invited us all to spend the night at their house, so I had another nice bed to sleep in. We sent out for pizza and rented a movie. The past two days made a great grand finale to one of the best vacations I've ever had.

Sunday, August 4—fly home to Denver. I'm sad it's over but glad to be home with my family again.

20

Baculite Mesa

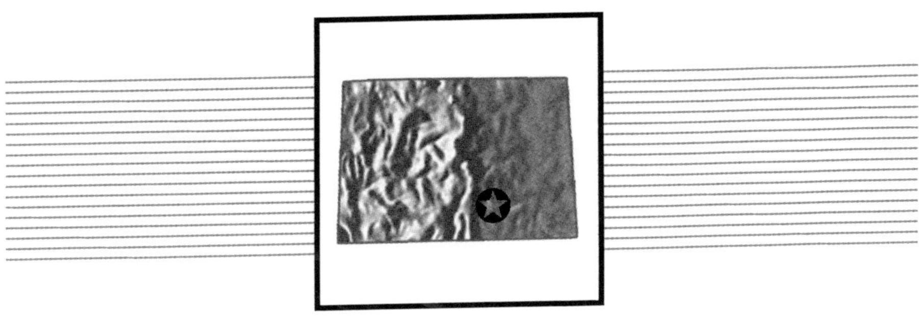

My friend Malcolm runs a field trip to Baculite Mesa, in southern Colorado, almost every spring. My mom and I had gone a few years ago, and found the site so prolific with the shells of these uncoiled ammonites that we came home with rock bags heavy with over 200 specimens after only a day of collecting. Most of what you find are casts, but occasionally you find one with some of the original pearly shell preserved, including the beautiful dendritic patterns of the complex sutures. In May of 1998, with

Baculite Mesa as seen from the south.

Mattie almost five and Allison eight, I decided that a trip to Baculite Mesa would be an ideal collecting opportunity for the kids. Allison was already a real dinosaur aficionado, and had a beautiful fossil leaf she found at Florissant three years before. Mattie was still too young to really be into fossils, but loved to collect things, and they had both had a good time hunting trilobites at Antelope Spring two summers before. I knew that a place with a lot of fossils and not too much hard walking or digging would make a great trip for the three of us.

Baculite Mesa is only about 5 miles east of the north end of Pueblo. The drive from Boulder takes about 3 hours, and none of us was the least bit thrilled with the idea of getting up early to be down there by 9, so we drove down Friday night and got a motel. (Getting the chance to stay at a motel, one with a swimming pool no less, excited the kids just as much as the prospect of finding fossils.) The site is on private land, so we had to have permission from the land-owners to collect. We assembled on top of the mesa Saturday morning, and Malcolm gave us an orientation as to the geology and pale-ontology of the region before we set out collecting.

Malcolm's collection of Pierre Shale fossils, many with the pearly nacre preserved.

Baculite Mesa exposes the Cretaceous Pierre (pronounced "peer") Shale, a marine unit be-tween 82 and 67 million years old. At that time, a great shallow ep-eiric seaway cut the continent of North America in half, connecting the Arctic Ocean with the Gulf of Mexico. The seaway was inhabited by mosasaurs, ammonites, and large fish, many of which are found in the chalk beds of Kansas laid down at the same time. In most places, the Pierre is not a very productive unit for fossils. It is a grey shale thought to repre-

Typical baculites from the mesa.

sent sedimentation in nearshore anoxic bottom waters, and thus does not preserve an extensive benthic fauna such as might be representative of more oxygenated conditions. Inoceramid clams are the exception, however, and are found throughout the Pierre. Baculite Mesa is a unique locality within the Pierre that is highly productive of the nektic (swimming), uncoiled ammonites called baculites.

Malcolm has been collecting the Pierre in the Baculite Mesa area for years, and showed us his nice collection of ammonites, baculites, clams, and snails, many with the pearly nacre and sutures preserved. He was careful to tell us, however, not to expect to find many such prime specimens in only a few hours, but he wanted to show us that the potential for some really nice finds was there. I hadn't seen anything like that the last time I was there, and I really felt that the kids would be more interested in finding a lot of stuff even if it wasn't so great, so we opted to collect on the flats near the south end of the mesa where we were assured that we would find lots of baculites.

Instead of having to scramble down the steep sides of the mesa like my mom and I had done, we now had the option of driving around the south end and parking on the flats, then approaching the mesa slopes from the south. That was great, because it made walking a lot easier, especially for the kids. Mattie had insisted on wearing her black patent-leather shoes, which are not exactly the best choice for tromping around between prickly pears, so we had to be careful. She was also wearing shorts, appropriate for the sunny, mid-70s weather, but which offered little protection against the cactus thorns and yuccas. But we started finding baculites almost right away. It only took a few minutes for me to show them what the baculites looked like, and then they began spotting them everywhere. The best places to look were where there wasn't too much ground cover, near the summits of the small, rolling hills. Our rock bags filled quickly, and soon I was the pack mule carrying three heavy rock bags, as the kids dumped more and more baculites into them.

Somehow or other the water bottle I had in my rock bag got left in the truck, and soon we were all getting thirsty. I was also getting tired of carrying those three heavy rock bags. So about midday we decided to call it quits and go back to the motel for a swim and a late lunch. We all had a really great time, and both girls were really proud of all their finds. Allison found several sections which had broken at

Mattie and Allison proudly show off their finds.

a suture, revealing the interfingering surface. Mattie was most excited when she found a "tail," or rounded end section. Both of them took their fossils to school for show and tell after we got home, and Mattie even used her baculites for her science fair exhibit the following year. Fossils don't always have to be big or showy to be fun to collect!

21

In the Footsteps of O. C. Marsh

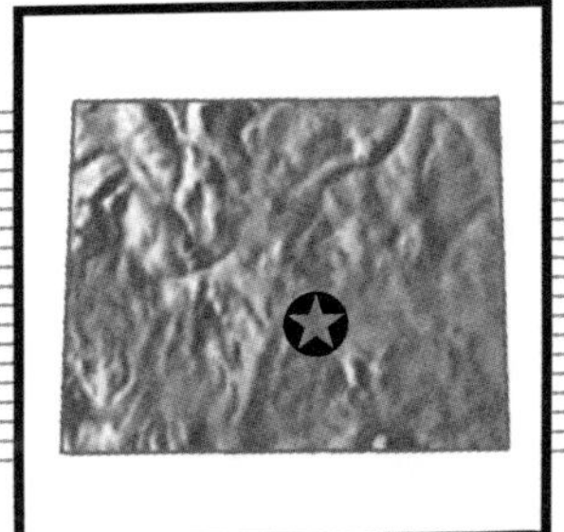

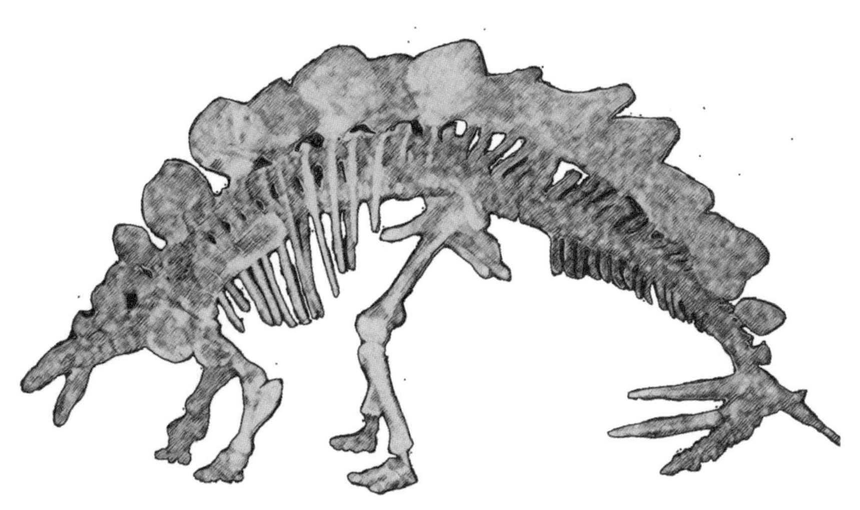

Most every dinosaur aficionado is familiar with the vehement rivalry during the late 19th century between the Yale University professor Othniel Charles Marsh and his arch-enemy, Edward Drinker Cope. Both scientists rushed to collect as many dinosaur bones as they could during the 1870s and 1880s, usually relying on field teams to do the actual collecting, and described dozens of (sometimes dubious) genera of the behemoths.

Como Bluff landscape.

One of Marsh's favorite prospecting areas was Como Bluff, Wyoming, in the Jurassic Morrison Formation. This formation, a terrestrial deposit laid down on low floodplains dotted with rivers and lakes, is a treasure trove of dinosaur bones and is widely exposed in the western states and provinces from New Mexico to Alberta. It has a very distinctive look to it, and once you learn it, you can recognize the Morrison just about anywhere, even from the freeway at 70 mph. It consists of alternating sandstones and mudstones that, in particular layers, have either a purplish or a grey-green cast. Whenever you see those colors exposed in the Western Interior, you know in an instant it's the Morrison.

The formation was named for the tiny Colorado town of Morrison, west of Denver and south of Golden, which was the site of Arthur Lakes' famous *Stegosaurus*, *Apatosaurus*, and *Diplodocus* discoveries of 1877. The area where those bones were found is protected today as Dinosaur Ridge, a National Natural Landmark, which also

My dad with Mattie at the Bone Cabin on a previous trip. The walls of this structure are constructed entirely of pieces of dinosaur bone.

includes Cretaceous exposures that preserve scores of dinosaur foot-prints laid down along the shoreline of the later Western Interior Seaway.

But back to Como Bluff. The area of high, shortgrass prairie and rolling hills is mostly ranchland today, and dinosaur bones are still to be found there. In July of 1998 I took Allison and Mattie up to Wyoming to help excavate a *Stegosaurus* which an acquaintance had found on a ranch near Medicine Bow, a one-horse town a ways north of Laramie.

There are a couple of hotels in Medicine Bow (the Trampas Lodge and the Virginian, for those of you who remember the old TV show), but we chose to camp out for free at the city park. It's a small block of emerald green in this buff-and-sage land, with no formal camping facilities, but it does have playground equipment and restrooms, as well as soft green grass on which to pitch your tent. After having an unpleasant experience with the automatic sprinklers coming on at 11 p.m. on a previous trip, soaking our tent and some of our bedding, this time we equipped ourselves with several buckets and concrete blocks so that we could effectively "shut off" the sprinkler heads close to our tent. We had the park to ourselves and the town is very quiet, so all in all, you couldn't beat it.

No trip to Como Bluff would be complete without stopping at the Bone Cabin Quarry Museum a little way out of town. This is the site where the old-time prospectors constructed a dwelling entirely out of "cobblestones" made from dinosaur fossils. The cabin stands today and is the home to a small museum of paleontological and his-torical artifacts. It doesn't have regular hours; you pay your small admission fee at the caretaker's house, and he unlocks the cabin for you and lets you look around.

The *Stegosaurus* dig was about 20 miles northeast of town on good gravel roads. Chris Weege, the guy who leases the "dinosaur

Quarry pit.

rights" to the ranchland, had constructed a lean-to shelter over the southern and western sides of the pit, to keep out the worst of the afternoon sun. Those of you who haven't been to Wyoming would think it is far enough north that heat wouldn't be a problem. Ha! Wyoming summers can be almost as intense as those in New Mexico or Arizona, and there isn't a stick for shade on the prairie. So some kind of makeshift shade canopy is an important part of quarry comfort.

The bones all lay in a grey, sandy-mudstone layer, within perhaps a 10' x 12' area. While some layers of the Morrison are quite soft, and digging can be easily accomplished with small tools, this was a distinctly consolidated layer—hammer-and-chisel territory. I had brought some tools of my own, but Chris also had extras for volunteer helpers, so the kids and I could all work at the same time. The bones were black against the grey matrix, and easy for even five-year-old Mattie to distinguish from the rest of the rock. Nevertheless, for the most part the kids worked on removing matrix from areas a few inches away from the bones, just in case.

The quarry floor sloped down to the east, and the three of us worked in the higher-up section, near the west wall of the shelter. Several other folks were also up there helping at the same time we were, and they worked the pit area, removing and jacketing bones from the disarticulated, jumbled-up dinosaur.

If you've never done any jacketing, you really ought to go on a dig sometime where you can participate in this process. The technique hasn't changed much in over a century, when the old-time prospectors devised a method of protecting delicate bones for transport by mule and rail to eastern museums. After the bone has been stabilized with Vinac and any cracks mended with superglue, it is coated with a layer of material to keep the jacket from sticking to the bone. Some people like to use aluminum foil, whereas others prefer toilet paper wetted down with dabs of a paintbrush. Either way, the objective is to make the jacket easy to get off the bone once it is back at the lab (or in this case, Chris' garage). After the separation layer is applied, strips of burlap dipped in plaster are added in a papier-maché fashion to the top surface of the bone. The bone's bottom surface is still embedded in the matrix at this point. The plaster is allowed to dry overnight, and the next day the rock is chiseled away from most of the underside of the bone, forming a mushroom-like pedestal. Then crowbars are inserted underneath, and—as quickly as possible so as not to break it—the bone is flipped upside down. Now burlap and plaster can be applied to the undersurface, forming a complete jacket encasing the bone. If it's a

Stegosaurus bones.

particularly large element, a piece of wood may be attached to the jacket to ensure that it doesn't flex and break during transport. Fortunately, nowadays, jacketed bones can be gotten back to civilization by pickup truck rather than by mule train.

Allison and Mattie excavating the bones.

We were only there for a week during July, but Chris had various people come up from Denver off and on during the summer. His garage is now practically filled with bones from not only that stegosaur but also an *Allosaurus* he found in the same area, enough to keep him in prep work for a long time to come. Since the bones come from private land, he's free to keep them, sell them, or donate them to a museum as he chooses. I don't see him often and haven't heard what fate he decided for our *Stegosaurus*, but the experience is one my daughters will not soon forget. How many kids their age have a mom who's loony enough to take them on a real dinosaur dig?

22

Flint Hill

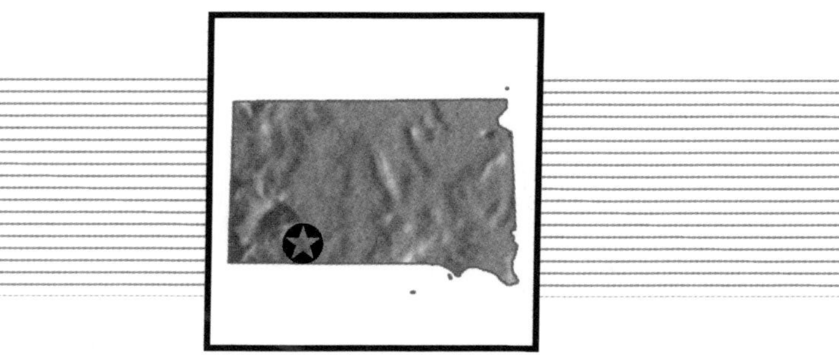

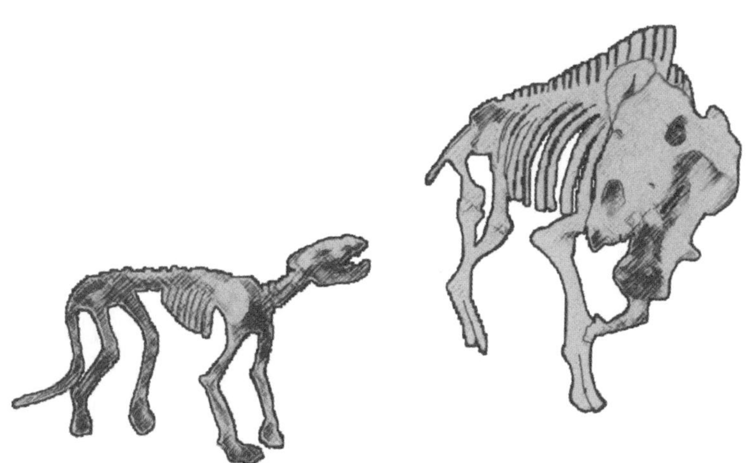

Just to show you I'm not the only nut who threw in the towel on a corporate career to go into the less lucrative field of paleontology...my friend Toni quit the phone company after 20 years to pursue her Master's at South Dakota School of Mines. In the summer of '00 she was finishing up the fieldwork for her thesis, and I drove up to the site a little west of Martin, South Dakota to assist her with the digging.

Looking towards the Little White River from Flint Hill.

I was a little apprehensive about going, because Angie, who had been there the previous year, had told me that it could get over 100° and that there was nothing bigger than a blade of grass for shade. I don't sweat a lot, and even if I dress lightly, wear a hat, and drink plenty of fluids, being out in the hot sun and getting overheated is a sure trigger for one of my killer migraines, but I decided to go anyway and hope for the best. I hadn't been on a mammal dig in 13 years, and those had been very different experiences—one a Pleistocene cave deposit, the other collecting microfossils from anthills in the Pawnee Grasslands.

Well, Toni must drive 80 mph and never make pit stops, because what she told me was a 6 hour drive from Denver took me nine and a half. It was getting dark when I approached the turnoff, only

The digs at Flint Hill: Regina in foreground, Toni with white hat.

Bone fragments I found by screening.

to discover that there was positively no road leading south from mile marker 143. After exploring a couple of long ranch driveways and cowpaths through cornfields in the general vicinity, I decided to look for a motel in Martin and try again in the morning. It even did occur to me that she might have transposed the digits on the directions, but I was too tired to want to deal with it in the dark, especially since the site was several miles in on four wheel drive roads punctuated by those horrible barbed wire ranch gates.

Sure enough, in the morning, right at mile marker 134 were the flags indicating the turnoff, and the landmark windmill a little further on. It was close to lunchtime when I found the Hill (I'm not an early riser, if I have my choice) and was greeted by Toni and the rest of the crew.

The site is in the Miocene Batesland Formation, a stream-deposited sandstone/mudstone unit around 15 million years old. The terrain was savannah during the Miocene, and the fauna consisted of a variety of ungulates (hoofed mammals) such as rhinos, gazelle-camels, and entelodonts (nightmarish, predatory pigs), as well as chalicotheres, which look like a cross between a horse and a giraffe, some carnivores and the usual "shadow creatures" like rodents and

Wolf and entelodont face off at Agate Fossil Beds.

snakes. As is generally the case with fluvial deposits, you don't find articulated skeletons, because the flowing water disperses the bones. But you do find a lot of single elements, and where the river overflowed onto the floodplain, you can even find small and delicate bones.

Being the tail end of the scheduled digging, most of the gridding and removal of big bones had already been done. Toni and Diane were measuring strat sections and Regina was doing surface surveys, so I chose to sift through the previously-removed sediment in search of small bones that had been missed the first time around. I wanted a task that wasn't too physically demanding since the weather was quite warm, and I wanted to avoid getting overheated.

Working alone suited me just fine, and I had a beautiful view across the valley of the cottonwoods lining the Little White River. There were a few clouds and a nice breeze to keep it from getting too hot, but not enough wind to blow a lot of my sifting dust into my face. Toni had some pre-built screens but the mesh on all of them was too big for the things I was looking for; so I made do just fine with a plastic tray and a piece of plastic window screen, just scooping up sediment with my trowel and scratching till all the dust went through, then examining what remained on the screen for small bones. Most of what I found was unidentifiable broken sections of longbones, a millimeter or so in diameter and a few long, but I did find a bird femur (thigh bone) and a nice rodent astragalus (ankle bone).

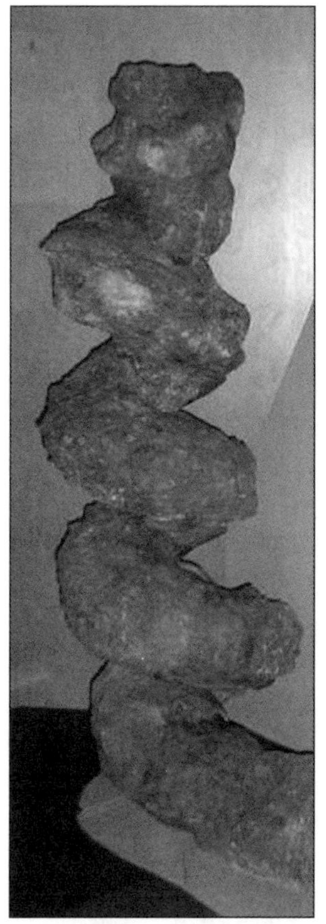

Daemonelix.

Unfortunately, the digital camera I had then didn't have a zoom, and even in macro mode, the pictures I took of the small fossils I found don't show much. They do, however, give you an idea of the types of residual material you end up with when you look for fossils by screening sediment.

By noon on Friday the partly cloudy weather had given way to the baking sun that Angie had warned me about. By midday I had that pressurized feeling in my head that lets me know I'd better get out of the sun fast or I'm going to be very sorry real soon, and since it was our last day of fieldwork anyway, I told Toni I thought I'd better pack up my gear and hit the road. Fortunately nowadays there are much better migraine drugs than there were a dozen years ago, and I was able to avoid a repeat of my Hungry Hollow experience. I decided to break the drive home into two days so that I'd have time to swing by Agate Fossil Beds National Monument in western Nebraska and see some good examples of similar, slightly older (20 mya) fossils. It was well worth the detour, as they have a very nice museum with impressive mounted skeletons of an entelodont, a towering chalicothere, and some smaller denizens of the Miocene savannah. I was particularly intrigued by the *Daemonelix* ("Devil's

Covered wagon at Scottsbluff.

corkscrew")—the fossilized burrows of extinct, land-dwelling beavers, something I'd read about but never seen before. I also stopped by Scottsbluff and soaked in a little lore about settlers travelling west in their covered wagons.

I arrived home just in time to catch a beautiful sunset behind what we in Boulder call the Flatirons, sandstone massifs that lie at a steep angle against the foothills, giving silent testimony to the ancestral Rocky Mountain range that preceded our current one by a quarter billion years—a fitting postscript to a delightful expedition.

Postscript

I hope you've enjoyed reading these tales of what fossil-hunting can be like in a variety of settings and under all sorts of conditions. Whereas there are a number of books on the market that include some measure of information on fieldwork, they usually recount stories of large, professional, well-financed expeditions to the exotic realms of Africa or Mongolia—things "regular folks" don't have a prayer of getting in on. My trips have all been things that I did on my own, or was able to participate in as a volunteer through WIPS or a university or museum. The vast majority of them didn't cost me a cent other than what I spent on transportation to and from there. And the experience they provided was priceless.

If this type of thing sounds like your "bag," opportunities abound if you know where to look for them. Find an amateur group in your area and join. They are usually your best resource for finding the good collecting spots nearby, hooking up with other fossil fans who know of digs or have friends who do, and in arranging cooperative ventures with state or national parks that have fossil resources. Another good source of opportunities is a natural history museum, whether public, private, or associated with a university. Through them, you can participate in digs that require professional supervision and a permit, such as for collecting vertebrate fossils on public lands. You won't be able to keep the fossils you find, but you'll gain experience that it is difficult to get anywhere else. Most also welcome volunteers who will work back in the lab doing preparation and curation work—something I've also done, but which is outside the scope of this book.

Don't even know how to locate these groups or institutions? Now that practically everyone is connected to the Internet, that's a good place to start. Earthwatch, the organization that sponsored my first dig at Fruita, is at http://www.earthwatch.org. Their expeditions are great, but some of them can be a little pricey. If that's not for you, either search on Google or go to the Open Directory Project at http://www.dmoz.org. It's a hierarchical directory of websites that is absolutely unparalleled. Start in the "science" category and work your way down through earth science and paleontology. Museums, groups, resources, publications—everything you're looking for will be right there. Just nose around until you find what you're interested in.

Happy hunting!